The Devonian tetrapod *Ichthyostega*

T0257369

ERIK JARVIK

Jarvik, E. 1996 04 15: The Devonian tetrapod *Ichthyostega. Fossils and Strata*, No. 40, 1–213. Oslo. ISSN 0300-9491. ISBN 82-00-37660-5.

The material of *Ichthyostega* Säve-Söderbergh 1932 collected by the Danish expeditions to East Greenland (1929–1955) led by Lauge Koch is described and documented by photographs of all important specimens. The ichthyostegids are related to the osteolepiforms, and comparisons with the Devonian osteolepiform fish *Eusthenopteron* and post-Devonian stegocephalians are made. In some respects the ichthyostegids are fish-like, whereas in others they have reached the tetrapod level. However, they also show specializations unparalleled in later tetrapods and belong to a sidebranch of the Osteolepipoda. Several new structures, such as canals for the occipital arteries, the prominent sacculus vesicle, the parotic crest of the palatoqudrate, the attachment areas for a paired basicranial muscle, and an arcual plate dorsal to the notochord are described. Special problems, e.g., the terminology of the dermal bones of the skull roof, cranial kinetism, resegmentation, the metapterygial stem, and the origin of the tetrapod limbs, are discussed. It is concluded that the hindlimb in *Ichthyostega* is pentadactyl and that pentadactyly is primitive for the Osteolepipoda. □*ICHTHYOSTEGA*, Tetrapoda, Osteolepiformes, Stegocephalia, anatomy, evolution, Devonian, Greenland.

Erik Jarvik, Department of Palaeozoology, Swedish Museum of Natural History, Box 50007, S-104 05 Stockholm, Sweden; 27th December, 1994; revised 3rd March, 1995.

Contents

Historical review and collecting work

In 1897, the Swedish engineer S.A. Andrée with two companions, starting from Spitsbergen, made an attempt to reach the North Pole with the balloon 'Örnen' (the Eagle). The attempt failed, and in the two next summers rescue expeditions were sent out. These expeditions were under the leadership of the Swedish palaeobotanist and geologist A.G. Nathorst, who in 1898 with his ship 'Antarctic' searched for the lost balloonists in the area of Spitsbergen and King Charles Land. Next summer he continued his searching in the coastal area of northeastern Greenland (Nathorst 1900). This part of Greenland is usually protected by a more or less broad belt of drift-ice, not easy to pass with a combined sail- and steam-powered ship as that used by Nathorst. However, in 1899 the conditions were favourable, and after ten days in the pack-ice the expedition members had eight weeks to explore this uninhabited and imperfectly known part of East Greenland. The cartographer of the expedition, P. Dusén, made a good map of the area (about 72°–73° 60' N). with many new geographical names (Ymer Ø with Dusén Fjord and Celsius Bjerg, Gauss Halvø, etc.; Nathorst 1900, Pl. 2). In connection with his geological investigations Nathorst (1901, p. 294) found some scales on the northern side of the outer part of Dusén Fjord, and on the opposite side of the fjord, that is on Celsius Bjerg (Fig. 1A), he found some plates of vertebrates.

The fossils were sent to the British palaeontologist A.S. Woodward, who (1900) referred the scales to *Holoptychius nobilissimus* Ag. and the plates to a new antiarch species, *Asterolepis incisa*. On the basis of the fossils he established the geological age to be Upper Devonian.

After Nathorst's visit nothing of interest to us happened in East Greenland until 1924–1925, when a new Eskimo colony was founded at the mouth of the mighty Scoresby Sund Fjord. Starting from that colony, the Danish geologist Dr. Lauge Koch (Fig. 1C) made two sledge journeys (October–November 1926, February–June 1927) northwards along the eastern coast of Greenland. During these journeys he made plans for a detailed exploration of this coastal area. He then paid special attention to the Devonian areas discovered by Nathorst, but because most of the rocks were snow-covered, he was unable to find any fossils. Remembering the fossils found by Nathorst, Dr. Koch for his first expedition by ship (1929) engaged a Swedish geologist, Dr. O. Kulling, for collecting and stratigraphical work in the Devonian (Koch 1930). Kulling was successful, and a great collection of vertebrates, mainly from the northern slope of Celsius Bjerg (Kulling 1930, 1931), was brought home. Dr. Koch then telegraphed to Professor Erik Stensiö in Stockholm and asked him if he was interested in describing Kulling's material. Stensiö answered in the affirmative, and in this way started an intimate cooperation between Lauge Koch and the Palaeozoological Department of the Swedish Museum of Natural History in Stockholm. At this time there was a struggle between Denmark and Norway for the hegemony of East Greenland, and in 1929 also a Norwegian expedition, led by the engineer A.K. Orvin, was working in the area of Celsius Bjerg. The material collected during this expedition (Orvin 1930) was rapidly described by Professor A. Heintz (1930) who, besides *Holoptychius* scales, identified two important Upper Devonian placoderm genera, *Bothriolepis* and *Phyllolepis*, each with one species (*B. groenlandica* and *P. orvini*). Another interesting form collected by Orvin in 1930, a little below the summit of Celsius Bjerg, was described by Heintz (1932) as *Groenlandaspis mirabilis*. Because of the strained relations to Norway, it was easy for Dr. Koch to raise money from the Danish Government and various other sources. Without economic problems he could then prepare the large 'Three Year Expedition' (1931–1934) with two ships ('Godthaab' and 'Gustav Holm') and (from 1932) seaplanes. He also generously supported the preparation of the material submitted to Professor Stensiö. In his description of this material, Stensiö (1931) referred *Asterolepis incisa* Woodward to the new genus *Remigolepis* and described several other species of that genus. Kulling's material also

Fig. 1. □A. Northern slope of Celsius Bjerg, Ymer Ø, East Greenland, where the first ichthyostegids were collected in 1929 (cf. Fig. 6). □B. Eastern part of southern slope of Sederholm Bjerg in Paralleldal. From Jarvik 1935, Pl.1:2 (cf. Fig. 6). □C. Lauge Koch (1892–1964). □D. Gunnar Säve-Söderbergh (1910–1948). *P*, Phyllolepis Series (Group); 1, Lower Red Division (Aina Dal Formation); 2, Middle Grey Division (Wiman Bjerg Formation); 3, Upper Reddish Division (Britta Dal Formation) of the Remigolepis Series (Group); 4, the course of Profile 4 with the prolific horizon at 1174 m;   ; the horizon with *Phyllolepis* and *Bothriolepis*.

Fig. 2. In 1943, when only parts of skulls were described, the Danish cartoonist Storm-Petersen published this anachronistic first 'restoration' of *Ichthyostega,* or 'den firbenede fisk' (the fourlegged fish), as it was called by Danish journalists.

included scales of *Holoptychius* and others referred to as indeterminable 'crossopterygian' remains, among them a skull-roof of *Eusthenodon* (Jarvik 1952, Pl. 17:1) and some cleithra and a clavicle of *Holoptychius* (Jarvik 1972, p. 123). Of greater interest to us are, however, the two first-found specimens of *Ichthyostega, viz* No. 99, 'scales of a fish-like vertebrate of uncertain affinities' (Stensiö 1931, Pl. 36), later identified as ribs (Pl. 41 herein), and No. 220 (an imperfect skull, Pl. 1:1, 2), which was not mentioned by Stensiö.

On the recommendation of Stensiö, a young Swedish student, Gunnar Säve-Söderbergh (Fig. 1D), was chosen by Dr. Koch to continue the collecting and stratigraphical work in the Devonian during the 'Three Year Expedition'. For that purpose special books were printed, each with 100 leaves with detachable numbered tags to be attached to the fossils in the field and with place for notes concerning localities and other information. Back in Stockholm new tags with that information, a statement that the material belongs to Grøn-lands Geologiske Undersøgelse (GGU), and museum num-bers (A. for amphibians and P. for fishes; see, e.g., Pls. 13:2, 47:3) were printed, and special catalogues were initiated. As to new numbers, see p. 19. In the first year (1931), Säve-Söderbergh, assisted by Eigil Nielsen and two other Danish students, devoted most of the time to collection in or close to the localities discovered by Nathorst and Kulling on the northern slope of Celsius Bjerg (Säve-Söderbergh 1932b).

Fig. 3. Erik Stensiö (to the left) with lunch-guests: Dr. F.A. Bather, Elsa Warburg, Agda Brasch, Erik Jarvik, Eigil Nielsen and Gunnar Säve-Söderbergh. Photo Aina Stensiö, June 1932.

Among the rich material brought home to Stockholm (A.1–A.14) were some imperfect skulls (Nos. A.1–A.7) which Säve-Söderbergh (1932a), after a careful preparatory work, identified as belonging to stegocephalians of a new type, the first tetrapods found in the Devonian and the oldest tetrapods known so far. Two genera, *Ichthyostega* and *Ichthyostegopsis,* were distinguished and were placed in the family Ichthyostegidae of the new order Ichthyostegalia (Säve-Söderbergh 1932a, p. 100).

The discovery of Devonian tetrapods caused a great sensation, and naturally it attracted attention in the Danish press. *Den firbenede Fisk* ('the four-legged fish'), as the animal was called by Danish journalists, became a hit in the fight for East Greenland, which in 1933 was adjudged Denmark (Decision of the International Court of Justice in the Hague). Somewhat later, the Danish cartoonist Storm-Petersen made a first 'restoration' of the whole animal (Fig. 2), a notable achievement since only parts of the skull were known.

In the spring of 1932, Säve-Söderbergh asked me if I, together with two other students from Uppsala (Fig. 5), would like to participate in the second year of the 'Three Year Expedition', the following summer. In order to be familiar with the kind of fossil material we were expected to find, I paid a short visit in June to the Swedish Museum of Natural History in Stockholm, where I had the pleasure of meeting Professor Erik Stensiö, the preparator Miss Agda Brasch and the Danish palaeontologist Eigil Nielsen for the first time (Fig. 3). I then went to Copenhagen and on June 16 started my first trip to East Greenland, this time on board 'Gustav Holm' (Fig. 4), a wooden ship of the same type as used by Nathorst. Some of us, at least, enjoyed the stormy crossing of the North Sea, and after 11 days at sea we reached Akureyri on the northern coast of Iceland. We had to stay there until July 5, and because of unfavourable ice conditions a further 11 days were lost until the ship (on July 17) could anchor outside the wintering station Eskimonæs on Clavering Ø. Unfortunately, the M/S Godthaab with Säve-Söderbergh on board was still tackling the drift-ice. Therefore, for a week I had to assist Eigil Nielsen collecting Triassic fishes and stegocephalians in the area of Kap Stosch. Finally, on July 27, Säve-Söderbergh could collect his party, which, supplied with a small motor-boat, was disembarked on the famous northern side of Celsius Bjerg (Fig. 1A). One of Säve-Söderbergh's tasks, and perhaps his main interest during the field-work, was stratigraphy. Interrupting the search for fossils in the old

Fig. 5. Our camp 1932 on the southern side of Celsius Bjerg at Sofia Sund. From the left: Gunnar Säve-Söderbergh, Gunnar Lindgren, Torgny Säve-Söderbergh and Erik Jarvik.

localities, we made excursions with the motor-boat. In this way the mountains (named Siksakbjerg, Teglbjerg and Blaskbjerg) surrounding the outer part of the Dusén Fjord (Fig. 6) were visited during two days, but very few fossils and no ichthyostegids were found. A longer trip to the southern side of Celsius Bjerg (Fig. 5), including a short visit to Rudbeck Bjerg on the southern side of Sofia Sund, in addition to numerous remains of *Remigolepis* and 'crossopterygians' yielded only two questionable ichthyostegid fragments (A.26, A.27). A profile was measured from the shore of Sofia Sund to the Vest Plateau of the western part of Celsius Bjerg (Säve-Söderbergh 1933b, p. 17). After a tiring climb to the top of the plateau (at 1296 m) Säve-Söderbergh spent hours in the cold photographing and making drawings of the surrounding mountains. When we, after 16 hours, returned to the camp with our rucksacks loaded with samples of rocks for planned sedimentological investigations but few fossils, he continued to make his notes long after his exhausted assistants had fallen asleep. This is an example of the energy, indefatigability and extraordinary working capacity that characterized Säve-Söderbergh both in the field and at home in the laboratory.

After our sojourn at Celsius Bjerg which (in addition to *Remigolepis,* 'crossopterygians' and lungfishes) resulted in several specimens of ichthyostegids (A.15–A.25, A28–A.32, A.220), we were transported to the shore of Gauss Halvø on the northern side of Kejser Franz Joseph Fjord (Fig. 6). Säve-Söderbergh started naming the coastal mountains after famous palaeontologists. From the west: Smith Woodward Bjerg, Stensiö Bjerg, Wiman Bjerg, and Nathorst Bjerg (Fig. 7, left). The valleys between the mountains were called Britta dal (after Britta Arnell, who later became his wife), Aina dal (after Mrs. Aina Stensiö), Elsa dal (after Dr. Elsa Warburg) and Agda Dal (after Miss Agda Brasch, who cleaned and prepared the fossils in Stockholm). In addition to great collections of *Remigolepis* and 'crossopterygians', several ichthyostegids (A.34–A.36, A.167, A.170, A.253, A.255), mainly

Fig. 4. Top left, middle, and bottom right: The expedition-ship Gustav Holm. Top left: Fastened with ice-anchor to the pack-ice close to the coast of East Greenland, 1932. Bottom right: Seen from behind with the open-cabin Heinkel hydroplane. Middle left: Forcing the pack-ice toward the nearby Greenland coast, 1932. Middle right: In full sail. Bottom left: Disembarkment in Kejser Franz Joseph Fjord at Nathorst Bjerg (1948) from Norseman hydroplane via motor-boat and ice-floes. Top right: Lauge Koch prepared for flying 1932 in an open-cabin hydroplane, dressed in a flying suit of polar bear fur.

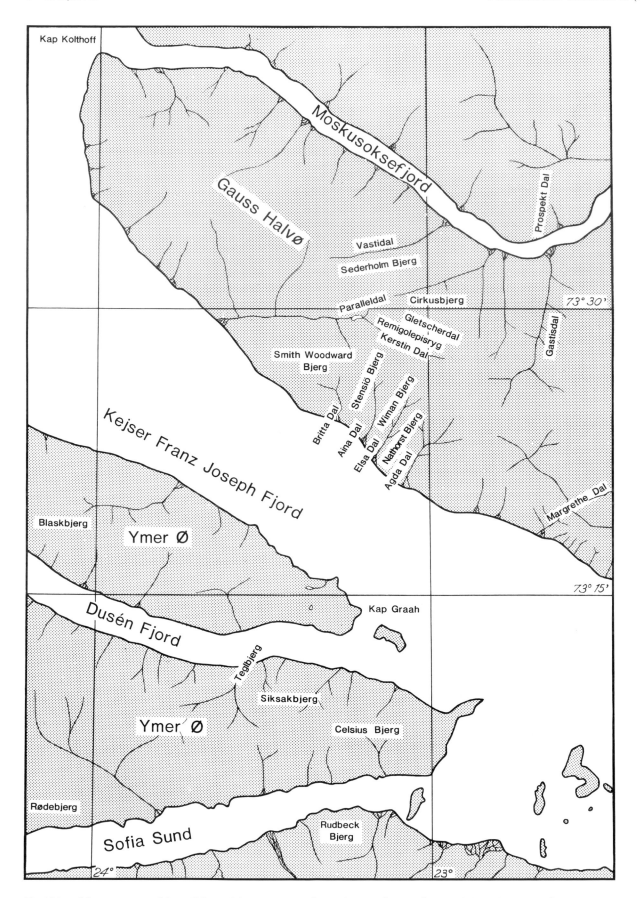

Fig. 6. Map of the western part of Gauss Halvø and the eastern part of Ymer Ø. From Sheet 73. Ö. 2. Geodetic Institute, Copenhagen, 1937.

Fig. 7. Left: Nathorst Bjerg with our camp in 1948. Right top and middle: On our way through the western Paralleldal to the camp in 1934 near the small lake at the col of the valley, with our five Icelandic horses loaded with camp outfit and wooden boxes with provisions. Right bottom: Loading the horses in the Phyllolepis Series with big blocks containing fossils packed in the wooden boxes for transport down to the Kejser Franz Joseph Fjord. Sederholm Bjerg in the background.

from the southern slope of Stensiö Bjerg, were secured. But the most remarkable find was a skull (A.33, Fig. 8) of a quite new type of tetrapod (*Acanthostega* Jarvik 1952) collected on Wiman Bjerg. I still remember Säve-Söderbergh sitting for hours in our tent, twisting and turning this specimen with a puzzled face.

In 1933 Säve-Söderbergh (1934b) devoted most of his time to stratigraphy. With two Swedish assistants he travelled in a motor-boat along the northern coast of Kejser Franz Joseph Fjord, from Margrethedal to Kap Kolthoff (Fig. 6). Large blocks with *Remigolepis* were secured on Nathorst Bjerg, but the most remarkable find there was a skull-roof of a large dipnoan (the holotype of *Soederberghia* Lehman 1959). Reaching the mouth of the Paralleldal, he walked alone through the western part of the valley up to the col. There he expected to find a party which, with three Icelandic horses from their camp at the Moskusokse Fjord, had planned to go through the E. Paralleldal to the col. However, the party, including Gustav Wängsjö and another Swedish student, had failed and Säve-Söderbergh had to walk back the long way (15 km) to one of his camps at Kejser Franz

Joseph Fjord. With the motor-boat and some of his assistants, he then rounded Kap Kolthoff and continued to the camp of the horse party at the delta in front of the Vastidal. With united efforts, his assistants succeeded in bringing the horses through the whole Paralleldal to Kejser Franz Joseph Fjord, while Säve-Söderbergh went back with the motor-boat. After a short rest he returned with all his assistants and the three horses to the col of the Paralleldal, where he camped for four days. Richly fossiliferous deposits with *Phyllolepis* and *Bothriolepis* were discovered at the base of Sederholm Bjerg (material described by Stensiö 1934, 1948), and a profile was measured to the top of that mountain (Fig. 1B). The mountains (Remigolepisryg and Smith Woodward Bjerg) on the southern side of the valley were also investigated. Only two rather poor ichthyostegid fragments (A.37 from Nathorst Bjerg and A.254 from Stensiö Bjerg) were found this year.

It may be added that in this year (1933) the Swiss geologist H.Bütler, as a member of Lauge Koch's expeditions, started his several years' stratigraphic and tectonic studies of the Devonian in East Greenland (cf. Bütler 1935).

Fig. 8. The type species of *Acanthostega, (A. gunnari)*, collected on Wiman Bjerg by Gunnar Säve-Söderbergh in 1934 (see Jarvik 1952 Pl. 21:1).

In 1934 Säve-Söderbergh (1935b, 1938) with two Swedish assistants travelled with a motor-boat to the inner part of the Moskusokse Fjord and devoted five days to investigate the mountains surrounding the Gästisdal (see below) and the Prospektdal (Fig. 6). The boat party then returned through the fjord and rounded Kap Kolthoff, and, passing the mouth of the W. Paralleldal, Säve-Söderbergh paid a short visit to our camp at the col of the valley. The boat party then passed over Kejser Franz Joseph Fjord to the northern side of Ymer Ø. However, on August 19, when the party was camping a little to the west of Kap Graah, the motor-boat was crushed by a furious gale. After a dramatic search, the boat-party was rescued on August 27, and taken on board the ship for the home journey.

In Gästisdal, Säve-Söderbergh discovered *Groenlandaspis mirabilis* (material described by Stensiö 1936, 1939) in strata which he earlier had called 'The upper Sandstone Complex', and as a final result of his stratigraphic investigations the Upper Devonian strata were divided as follows (Säve-Söderbergh 1935b, 1938; cf. Jarvik 1948b, 1950b, 1961):

Groenlandaspis Series (including the Upper Sandstone Complex and the Arthrodire Sandstone Series)

Remigolepis Series (in Paralleldal divided in an upper reddish [R3], a middle grey unfossiliferous [R2] and a lower dark-red [R1] division)

Phyllolepis Series (Lower Sandstone Complex)

Ichthyostegids have been found only in the Remigolepis Series, mainly in the upper reddish, but also in the lower dark-red division. There they occur together with the last representatives of *Phyllolepis* and *Bothriolepis* (*P. nielseni, B. nielseni;* Stensiö 1939, 1948; Jarvik 1950b, p. 16, footnote 1).

In 1934, the boat party did not find any ichthyostegids. A horse party, which beside myself included G. Wängsjö, two young Danes, five Icelandic horses and an Icelandic horse-handler, was more successful. Lauge Koch, taught by bad experiences from previous years, had realised that it was necessary to take special care of the horses. Starting from Kejser Franz Joseph Fjord, we walked with our five horses, loaded with camp-outfit and wooden cases with provisions, through the W. Paralleldal (Fig. 7, right), to near the small lake at the col where we camped from August 2 to 22. The collection was made by G. Wängsjö, one of the Danes and myself, while the other Dane and the Icelander had plenty to do transporting fossils down to the fjord and provisions up to the camp. We continued the collecting work in the locality in the Phyllolepis Series on Sederholm Bjerg discovered by Säve-Söderbergh, and big blocks with fossils could now be loaded directly on the horses and transported to the shore (material described by Stensiö 1936, 1948). The southern slope of that mountain was searched for fossils, and two profiles were measured (Jarvik 1935). In one of them (profile 4), at a height of 1174 m, (Fig. 1B) we were happy to encounter a richly fossiliferous stratum, which, besides *Remigolepis* and 'crossopterygians' (*Holoptychius* and *Eusthenodon*), yielded numerous valuable ichthyostegids. This locality, visited several times, is the most prolific locality for ichthyostegids found by us; many specimens figured in this volume come from this site. Also the slopes of Cirkusbjerg, Remigolepisryg and Smith Woodward Bjerg were investigated, and three profiles were measured. The ichthyostegid specimens collected by us in 1934 include Nos A.38–A.67. The most remarkable among these is a fine skull (A.64, Pls. 3, 4, 10) of *Ichthyostega* with counterpart (Pl. 2:2), collected in Kerstin Dal and on our home voyage exhibited in the saloon of 'Gustav Holm' lying on a blue velvet pillow.

In 1936, from July 26 to August 25, Säve-Söderbergh (1937) made detailed stratigraphical investigations in the area of the Paralleldal and found several ichthyostegids (A.68–A.84, A.221). No less than 18 profiles on the southern slope of Sederholm Bjerg (Fig. 1B) were measured, and in the upper division (R.3) each stratum was given a special nomenclature (Ra1, etc). His intention was to make a detailed stratigraphy of the Remigolepis Series on the basis of the changes in the genus *Remigolepis* during the time the sediments of this series, about 670 m in thickness, were deposited. However, because of unforeseen and unfortunate circumstances, his plans could not be fulfilled. The large collections of *Remigolepis* made by Lauge Koch's expeditions from 1932 to 1955 are still undescribed, and the stratigraphical manuscripts (Säve-Söderbergh 1935b, 1937) dealing with the results from 1934 and 1936 were never published.

In the summer of 1937, when Säve-Söderbergh was doing his military service, the news reached him that he had been appointed to be Professor of Palaeontology and Historical Geology at the University of Uppsala. However, late in the autumn that year after he had moved from Stockholm to Uppsala and was preparing for his new duties, he was suddenly struck by a serious lung disease which, disregarding a period in 1942, kept him in bed all the time until his untimely death in 1948. Because of his illness he could not take part in the palaeontological English–Norwegian–Swedish expedition to Spitsbergen in 1939 led by Erik Stensiö. However, during his long stay in hospital, Säve-Söderbergh continued his scientific work, and several papers were published (Säve-Söderbergh 1941, 1945, 1946). In 1942 he seemed to be on the road to recovery, and on May 5 he gave a lecture in the Royal Society of Sciences at Uppsala on fossil lungfishes (published posthumously, as was another paper on lungfishes written before his illness) and on December 12, 1942, he attended the oral defence of my doctoral thesis. This was his last public appearance.

In 1946, G. Wängsjö and I made an excursion to the Hornelen field (Middle Devonian) in western Norway with, as far as fossil vertebrates are concerned, bad results. The first Danish post-war expedition to East Greenland under the leadership of Dr. Lauge Koch was in 1947, when I, together with a Danish assistant, searched for fossils on the southern side of Celsius Bjerg. We found valuable material of *Holoptychius* and *Eusthenodon* (Jarvik 1952, 1972) and an impression of the skull roof (A.85) of *Acanthostega*, but few other tetrapod remains (A.86–A.88).

On June 8, 1948, Säve-Söderbergh died only 38 years old and so did not live to witness the surprising finds of the postcranial skeleton of *Ichthyostega* made during the expeditions in 1948 and 1949. In 1948 two parties were engaged in the collection, one with G. Wängsjö and two Swedish assistants, and another with myself and one French and one Swedish assistant. From July 22 to August 10 we were working together on the northern slope of Celsius Bjerg.

In the rich ichthyostegid material (A.94–A.105, A.109, A.168, A.171) secured, there was a block (A.109), which, when in the autumn it was split by E. Stensiö and prepared by Miss A. Brasch, turned out to contain a tail and a well-preserved hindlimb (Pls. 35:1–3; 38; 39:1, 65:1, 2; 66) of *Ichthyostega*. From August 11 to September 2 we were collecting on the slopes of the mountains along the southern coast of Gauss Halvø. In the ichthyostegid material found there (A.89–A.93, A.106–A.108, A.110–A.115, A.169), specimen A.115, collected by Wängsjö's party on Smith Woodward Bjerg, is of particular interest. In this specimen (Pls.

21:5; 42; 45; 46; 53; 54; 68:5), which shows parts of the head and trunk, Miss Brasch could display the shoulder girdles and, for the first time, the humerus, the radius and the ulna, all perfectly preserved.

Another event in 1948 may be mentioned. When on our home voyage through Kong Oscar Fjord we passed Mesters Vig, some geologists entered the ship. With a contented mien they dashed a heavy rucksack containing a big lump of native lead down on the deck. This resulted in the opening of a lead-mine and the construction of a landing field for aeroplanes at Mesters Vig.

In 1949, G. Wängsjö, with two Swedish assistants, secured valuable ichthyostegid material (A.116–A.166) in the area of the Paralleldal. Most of the material (A.132–A.144, A.149–A.166), comes from the locality at 1174 m in Sederholm Bjerg (Fig. 1B) discovered in 1934, among them two (A.156, A.157) showing the tail.

In 1950, G. Wängsjö and H. Bütler collected Middle Devonian osteolepids in the area of Kap Franklin (Jarvik 1985, pp. 13–32). In the last half of July 1951, together with two Swedish assistants, I searched for fossils on both sides of Sofia Sund (Rødebjerg, Rudbeck Bjerg and the southern side of Celsius Bjerg, Fig. 6) with, as far as ichthyostegids are concerned, negative results. Early in the morning of August 1 we were brought to Maria Ø, and then transported by ship to the mouth of the W. Paralleldal where we were put ashore on August 3. Our party was now supplemented with two Icelandic horses and a Dane employed to take care of them. Unfortunately, the horses were not fully broken, but after some trouble we managed to load one of them and on August 4 we could start our 5 hours' march up to the col of the valley where we camped for 20 days. During our stay there, one of my assistants and the Dane using the single horse, transported fossils down to the fjord and boxes with provisions back to the camp. The collection was therefore made mainly by two of us. We searched on the slope of Sederholm Bjerg and collected some additional ichthyostegids (A.172–A183) in the locality at 1174 m (this summer visited three times). We also spent several days on Smith Woodward Bjerg and found a rich locality at 980 m in the lower red division (R1) of the Remigolepis Series (A.189–A.205). In addition to ichthyostegids (A.172–A.219) and *Remigolepis,* material of *Eusthenodon* (Jarvik 1985) and *Holoptychius* (Jarvik 1972) was secured. In 1952 and 1953 (and the following years up to 1958) H. Bütler continued his stratigraphical investigations in the Devonian. The fossil material collected by him and his assistants was sent to Stockholm, where I made the preliminary determinations. No traces of ichthyostegids were found in this material (as regards osteolepiforms, see Jarvik 1985).

Our last two visits to the Upper Devonian of East Greenland were in 1954 and 1955. In 1954, when I was assisted by two Swedish students, ice conditions in the fjords were unfavourable for landing with seaplanes, and we had to spend several days waiting for flights. On July 23 we started

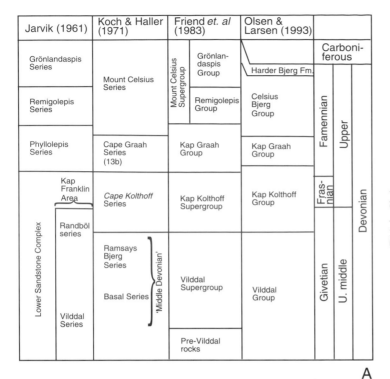

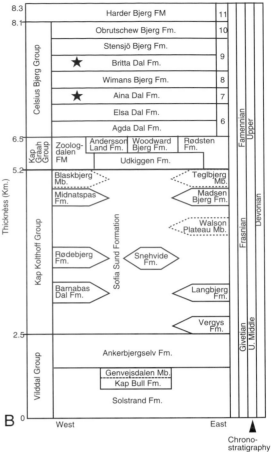

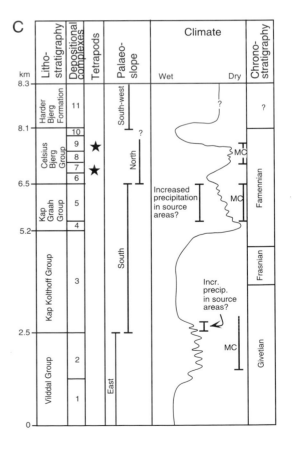

Fig. 9. □A. Stratigraphic schemes of the East Greenland Devonian, mainly to show correlation between subdivisions of the Famennian age deposits as based on (a) biostratigraphy as introduced by Säve-Söderbergh (1934a; subsequently revised by Jarvik); (b) allostratigraphy according to Koch & Haller (1971); mixed litho- and biostratigraphy proposed by Friend *et al.* (1976, 1983); and (c) the pure lithostratigraphy recently introduced by Olsen & Larsen (1993, Fig. 4). □B. Current lithostratigraphic subdivision of the East Greenland Devonian deposits. The numbers 6–11 refer to the correspondingly numbered depositional complexes of the Celsius Bjerg Group and the Harder Bjerg Formation in C. Occurences of tetrapods indicated by asterisks. Modified from Olsen 1993 (Fig. 2). □C. Schematic representation of the depositional complexes (1–11; cf. B:6–11) and their relationships to tectonically caused reversals of palaeoslope and to climatic conditions within the East Greenland Devonian basin. Occurrences of tetrapods in the Aina Dal Formation (7) and Britta Dal Formation (lower part of 9) appear to coincide with specific stages during increasing and subsequent decreasing aridity of the climate in late Famennian times when reversal of the basin palaeoslope resulted in northward-directed drainage system. Adapted from Olsen 1993 (Fig. 31), who discussed the cyclic aridity patterns (MC), also in the context of Milankovitch cyclicity.

Prepared by Dr. S.E. Bendix-Almgreen of Copenhagen who, besides Professor K.S.W. Campbell of Canberra, has refereed this paper.

our collection in the old sites on the northern slope of Celsius Bjerg and were on August 7 transported to the area of Kap Graah, where we had to stay until August 14. Our next goal was a locality with an interesting rhizodontid (*Spodichthys* Jarvik 1985) discovered by Bütler in 1952 on the northern side of Moskusokse Fjord, opposite the mouth of the Vastidal (Fig. 6). On August 21 we were taken away and flown to the southern side of Celsius Bjerg, where we stayed until September 1. In addition to specimens of *Phyllolepis*, lungfishes, *Holoptychius* and *Spodichthys*, the collection this summer also included several specimens of ichthyostegids (A.224–A.231).

In 1955, when I was assisted by H.C. Bjerring and another Danish student, we were dogged by much fog, rain and snow, which made collecting difficult. From July 27 to August 7 we camped at the mouth of the Britta dal on Gauss Halvø, for some days together with Eigil Nielsen and his assistants, and found some ichthyostegids (A.232, A.239–A.241, A.243) on Smith Woodward Bjerg and Stensiö Bjerg. When Nielsen's party had been transported to the area of Kap Franklin (see Jarvik 1985, p. 13) we travelled with motor-boat into the Moskusokse Fjord and camped at the mouth of the Vastidal. For three days we searched for fossils in that valley but, like Wängsjö in 1933, we found very little (*Remigolepis* and scales of *Holoptychius*). On August 10 our camp was therefore moved to the mouth of the E. Paralleldal. From there we walked with a part of our camp outfit packed in heavy rucksacks up through the valley, and after a seven-hours march we could pitch our tent opposite Cirkus Bjerg, about 9 km from the fjord. We made a futile excursion into the Gletscherdal, and the collection was restricted to the southern slope of Sederholm Bjerg, including a last visit to the locality at 1174 m, where some ichthyostegids were secured (A.233–A.238) and, among other things, some specimens of *Phyllolepis nielseni* from R1. On August 19 we walked back to the Moskusokse fjord and were on August 21 transported by motor-boat to the northern side of Celsius Bjerg, where we arrived on August 22. During our stay here (until August 31) we found a block with well-preserved ichthyostegid remains (A.248–A.251), some other ichthyostegids (A.244–A.247), lungfishes, an interesting skull roof (Jarvik 1985, Fig. 27B) and other remains of *Eusthenodon*.

In 1956, my eighth and last summer in East Greenland, I was assisted by two now well-known vertebrate paleontologists, S.E. Bendix-Almgreen and H.C. Bjerring. On July 27 our party was transported by helicopter to the southern Middle Devonian area in Canning Land. There we landed on a plateau on Hesteskoen, about 300 m above sea level and 3 km from Nathorst Fjord. The Middle Devonian rocks resemble those in the Upper Devonian, but our hope of finding tetrapods in these layers came to nought, and the search for tracks on the many flat rocks with ripple marks exposed in this area was also in vain. After collecting some material of osteolepids (Jarvik 1985), porolepiforms and placoderms, Bjerring, on August 18, was happy to find a richly fossiliferous stratum on the southeastern slope of Hesteskoen. This stratum yielded a good specimen of an osteolepid (*Gyroptychius dolichotatus* Jarvik 1985) but of greater importance, several more or less complete specimens of the porolepiform *Glyptolepis groenlandica* (Jarvik 1972). On August 29 a seaplane dropped a message informing us that the helicopters were out of order and that we had to carry all our fossils down to the coast of the icy Nathorst Fjord. After several tours up to our camp and back again we succeeded in bringing our heavy cases with the valuable fossils down to the shore, and on August 30 we were safely in Mesters Vig.

In 1958, H. Bütler finished his stratigraphic investigations in the Devonian of East Greenland (Bütler 1959, 1961). However, in the summers of 1968, 1969 and 1970, members of the Cambridge East Greenland Expedition, led by Dr. P.F. Friend, continued the stratigraphic work and carried out sedimentologic investigations in the Devonian. This resulted in a new division of the Upper Devonian strata (Fig. 9).

During these expeditions, some ichthyostegid remains were secured (Friend *et al.* 1976), but of greater interest are some pieces of rock collected on the southeastern slope of Stensiö Bjerg. Several years later Dr. Jennifer Clack discovered that a number of the pieces fit together to form one block which, to her great surprise, displayed cranial and other remains of several specimens of *Acanthostega* (Clack 1988a). Stimulated by this discovery, she contacted Dr. S.E. Bendix-Almgreen in Copenhagen, and in the summer of 1987 a British–Danish expedition, in the first place to Stensiö Bjerg, was arranged (Bendix-Almgreen *et al.* 1988, 1990). The collecting turned out successful, and a valuable material of acanthostegids and ichthyostegids was secured. This material, now in Cambridge, is being described by J. Clack.

Previous investigations

The material of fossil vertebrates collected by Säve-Söderbergh in 1931 on the northern slope of Celsius Bjerg (Fig. 1A) includes skull remains of fourteen ichthyostegid specimens (A.1–A.14). One example (A.2) excepted, they were collected on or in the talus cones above the East Plateau. Above the talus cones is a high and steep rock-face, and it is therefore not possible to say if the fossils come from one or more geological horizons. Only seven of the specimens were used by Säve-Söderbergh in his preliminary description of the ichthyostegids (1932a). These specimens were described as follows: A.1, *Ichthyostega stensiöi* n.sp. (type species; Fig. 10); A.2, *Ichthyostega?* sp. b; A.3, *Ichthyostega watsoni* n.sp; A.4, *Ichthyostega eigili* n.sp; A.5, *Ichthyostegopsis wimani* (type species); A.6, Ichthyostegid sp. a and A.7, *Ichthyostega? kochi* n.sp.

After preparation and studies of the fossil material, Säve-Söderbergh could make out the pattern of the dermal bones of the skull table, and, thanks to admirable preparatory work, he also could describe essential parts of the sensory canals. In order to be able to apply a reliable terminology he found it necessary to be familiar with the arrangement of the dermal bones and the course of the sensory lines in the presumed ancestral fishes, the imperfectly known crossopterygians and the dipnoans. He therefore prepared a description of the external dermal bones and the sensory canals of the Middle Devonian rhipidistid crossopterygian *Osteolepis macrolepidotus*. This study resulted in a publication (Säve-Söderbergh 1933a) which included a discussion of the terminology of the various parts of the sensory line system, and the relations between sensory canals and dermal bones. On the basis of the

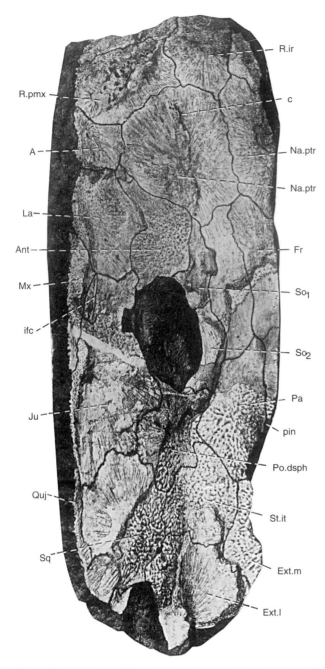

Fig. 10. The type species of *Ichthyostega, I. stensioei*. Imperfect cranial roof exposed from the ventral side; ×0.8. Abbreviations and terminology from Säve-Söderbergh 1932a, Pl. 1. *A*, anterior antorbital; *Ant*, posterior antorbital; *Ext.l*, lateral extrascapular; *Ext.m*, parieto-extrascapular; *Fr*, frontal; *Ju*, jugal; *La*, lachrymal; *Mx*, maxillary; *Na.ptr*, naso-postrostral; *Pa*, fronto-parietal; *Po.dsph*, postorbito-dermosphenotic; *Quj*, quadratojugal; *R.ir*, rostro-interrostral; *R.pmx*, rostro-premaxillary; *So1*, anterior supraorbital; *So2* posterior supraorbital; *Sq*, squamosal; *St.it*, supratemporo-intertemporal; *c*, canals, probably for vessels and nerves to the supraorbital sensory canal; *ifc*, infraorbital sensory canal; *pin*, pineal foramen.

new evidence, a restoration made by Stensiö of the skull roof in *Dictyonosteus*, a skull roof of *Eusthenopteron* with two pairs of parietals, and restorations of *Dipterus* by Stensiö and D.M.S. Watson, Säve-Söderbergh presented the interpretations of the dermal bones and sensory canals in the ichthyo-

stegids shown in Fig. 11. He also made comparisons with certain post-Devonian stegocephalians and demonstrated great similarities (in particular in the cheek) indicating relationship between 'rhipidistids', ichthyostegids and post-Devonian tetrapods. He also described the dermal bones of the palate and identified the external and internal nares. The unusual ventral position of the 'external nares' caused him some trouble, but after comparisons with *Dipterus* and (Säve-Söderbergh 1932a, Fig. 17) two 'indeterminable Crossopterygians' from the Lower Devonian of Spitsbergen with two external nares (anterior and posterior) he concluded (Säve-Söderbergh 1932a, p. 92) 'that it is the homologue of the posterior of the external nares that is found in the Ichthyostegids, whereas the more anterior external naris is not present in that group'. He also concluded (p. 98) 'that the external nares of the Dipnoans, like those of the Ichthyostegids are primarily situated on the ventral side of the snout'. This forced him to the conclusion (p. 99) that the external nares in later stegocephalians have secondarily migrated to the dorsal side of the snout.

In later publications Säve-Söderbergh (1934a, 1935a, 1936, 1941) referred to the ichthyostegids but did not add anything of importance. It may be mentioned, however, that in his 1934a paper he founded a new taxonomic unit, the Choanata, including dipnoans, crossopterygians and tetrapods. Among the tetrapods the ichthyostegids, the labyrinthodonts and the anurans (but not the urodeles) were classified as Batrachomorpha, whereas the anthracosaurs and the reptiles were united as the Reptilomorpha. In the latter taxon, although with some hesitation (Säve-Söderbergh 1935a, p. 202), he included also the birds and mammals. The Batrachomorpha and the Reptilomorpha were referred to a new taxon, the Eutetrapoda. The urodeles and the dipnoans, which in Säve-Söderbergh's opinion shared a common ancestry among the Choanata, were placed (Säve-Söderbergh 1935a) in separate classes (Dipnoi and Urodela).

In my paper on the snout (Jarvik 1942), detailed comparisons with ichthyostegids, other stegocephalians and primitive reptiles were included. In certain respects I could confirm Säve-Söderbergh's opinions, whereas in others I came to different conclusions. It was shown that the 'rhipidistids' include two distinct groups of choanate fish, one, the Porolepiformes, related to the Urodela and another, the Osteolepiformes, to the Anura. The 'indeterminable Crossopterygians' referred to by Säve-Söderbergh are porolepiforms, and *Dictyonosteus*, held to be a 'rhipidistid', turned out to be a coelacanthid. Stensiö's restoration of the skull roof used by Säve-Söderbergh is, as Stensiö pointed out to me and I could confirm, erroneous in several important respects (Jarvik 1942, p. 580). Since it was shown that the dipnoans lack choana, and this is true also of coelacanths, the term 'Choanata' was rejected. However, Säve-Söderbergh's opinion (see also 1936, p. 168) that the urodeles hold an isolated position among the tetrapods was confirmed, although they are related to porolepiforms and not to dipnoans as Säve-Söder-

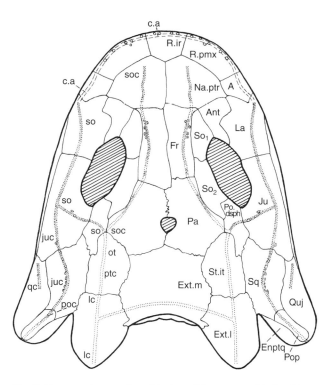

Fig. 11. 'Attempted restoration of the lateral line system of the head in an ichthyostegid. *c.a,* anterior commissure of the infraorbital canals; *juc,* jugal canal; *lc,* cephalic division of main lateral line canal; *ot,* otic part of the infraorbital canal; *poc,* preopercular canal; *ptc,* postotic canal; *qc,* quadrato-jugal canal; *so,* suborbital part of the infraorbital canal; *soc,* supraorbital canal.' For further explanations, see Fig 10. From Säve-Söderbergh 1932a, Fig. 15.

bergh believed. It was also shown that the ichthyostegids among the 'rhipidistids' agree with the osteolepiforms, and I could confirm that they are related to labyrinthodonts and anurans and that the Eutetrapoda is a reliable taxonomic unit. I could also confirm that the ichthyostegids belong to a side-branch and stated (Jarvik 1942, p. 640) that the ichthyostegids 'became specialized quite early in a way indicating that they cannot have given rise to any later Tetrapods known to us'. In this connection it may be worth being reminded of the surprising discovery that many of the Triassic labyrinthodonts (Jarvik 1942, pp. 632–635, Fig. 87; 1980b, Fig. 137) in important respects agree more closely with the osteolepiforms than with the ichthyostegids and other Palaeozoic tetrapods (Fig. 28). Another important specialisation concerns the external nostril (1942, p. 639) which, in contrast to Säve-Söderbergh's view, 'was an anterior external nostril homologous to the anterior external nostril in Crossopterygians and Tetrapods. Its ventral position must be secondary and is due to a downward migration from a primarily lateral position'. This has caused a reduction of the lateral rostral and other modifications at the margin of the mouth.

In a paper on the trigeminus musculature and in a review of vertebrate palaeontology in Sweden, Säve-Söderbergh

(1945, pp. 38–39; 1946, p. 363) briefly commented on my new ideas, which he in the main accepted. However, he still adhered to the view that the dipnoans are closer to the Crossopterygii, in particular the Porolepiformes, than to any other fishes.

At this time, in the middle of the 1940s, Säve-Söderbergh understood that he, because of his illness, would be unable to complete the taxonomic revision of the Scottish osteolepids. He had planned this as a preparatory work for the description of the material from the Middle Devonian of East Greenland collected in 1934 and 1936. The description of this material and of the large collections of Scottish osteolepids borrowed by Säve-Söderbergh from various museums was therefore committed to me, a work which kept me busy for three years. When the paper on the Scottish material (Jarvik 1948a) was accepted for printing, at Säve-Söderbergh's request, I wrote a note on the geological age of the ichthyostegid-bearing deposits. In that note (Jarvik 1948b) and in later papers (Jarvik 1950b, 1961) I could confirm Säve-Söderbergh's view that the deposits are upper Upper Devonian (cf. Spjeldnæs 1982, pp. 327–328). To capitalise on the work on the Scottish osteolepids, I then proceeded with the description of the Greenland osteolepids (Jarvik 1950a, b), some other material from Norway, and Upper Devonian osteolepiforms from Scotland.

Because of Säve-Söderbergh's long illness and his other tasks, the description of the ichthyostegid material had been at a standstill since 1932. After Säve-Söderbergh's death in June, 1948, Lauge Koch and Erik Stensiö decided that the description of this material should be entrusted to a person familiar with the ancestral fishes, and I was honoured with this task. At this time I was in the middle of my osteolepiform studies, and it was not until my last paper had been accepted for printing on May 10, 1950, that I was free to start my work on the ichthyostegids. However, before he was taken ill, late in 1937, Säve-Söderbergh had prepared some of the specimens collected in 1932 and 1934, and during his long illness and after his death the preparation was continued with skill and patience by Miss Agda Brasch under my supervision. This is true also of the extraordinary material of the postcranial skeleton found by the expeditions in 1948 and 1949. Therefore, by the middle of 1951, I was able to finish a description of the tail, which was provided with a fin like that in fishes but, in contrast to all later tetrapods, supported by special endoskeletal fin supports and dermal fin rays (lepidotrichia). In that paper (Jarvik 1952), parts of the pelvic girdle, the posterior part of the vertebral column, and postsacral ribs in *Ichthyostega* were also described. Moreover, a well-preserved hindlimb was figured, and the main elements were identified and named. However, since it had been established that the ichthyostegids are related to the osteolepiforms, I found it necessary to include, for comparative purposes, a description of the vertebral column, the ribs, and the median fin supports in *Eusthenopteron* (Fig. 38). The outstanding specimen No. P.222 and other excellent material were at my disposal for that purpose. These studies demonstrated the close similarity between the vertebrae of *Ichthyostega* and *Eusthenopteron*. In this connection may be mentioned that I also described the vertebrae in two dipnoans from the Upper Upper Devonian in East Greenland (*Jarvikia* Lehman 1959 and *Soederberghia* Lehman 1959), and it was stated that the vertebrae in the early dipnoans differ fundamentally from those in *Ichthyostega* and *Eusthenopteron*. After comparisons with *Eusthenopteron* and consideration of dipnoans and other fishes it was suggested that the tail fin in *Ichthyostega* includes equivalents to the second dorsal and caudal fins of the osteolepiforms (Fig. 38). According to this view, the second dorsal fin has fused with the epichordal lobe of the caudal fin, and the hypochordal lobe has been modified in a way indicating that the tail was dragged on the ground when the animal was walking. It was also suggested that the posterior process of the iliac portion of the pelvic bone was formed by incorporation of a postsacral rib (cf. p. 56).

In addition, a new large osteolepiform, *Eusthenodon wängsjöi*, was described, and the skull from Wiman Bjerg that had puzzled Gunnar Säve-Söderbergh, together with an incomplete skull roof found in 1947, was referred to the new genus *Acanthostega* and described as *A. gunnari*. The new genus was, with hesitation, assigned to the Ichthyostegalia.

Furthermore, an incomplete skull roof of *Ichthyostega* was figured (Jarvik 1952, Fig. 32), and new restorations of the skull table and the palate were presented. A vestigial subopercular was discovered, and my view (1942, p. 622) that the ventral migration of the external nostril had led to modifications of the lateral rostral was confirmed. As far as known, the lateral rostral is represented only by the sensory-canal component (Fig. 29). Another consequence of the ventral migration is that the anterior end of the maxilla has been driven inwards. The upper outer dental arcade formed by the maxilla and premaxilla therefore describes a curve inside the external nostril in a way unparalleled in later tetrapods. This unique specialization further debars the ichthyostegids from close relationship to later tetrapods and emphasizes their isolated position.

The grinding series of the excellent skull of specimen P.222 of *Eusthenopteron foordi*, started by E. Stensiö late in the 1920s, had been completed in 1951. When my paper on the fish-like tail of *Ichtyostega* was accepted for printing, work on the construction of the wax models on the basis of the serial drawings took a high priority. When this work was finished in 1952, I had at my disposal a unique material including not only the serial drawings and wax models but also numerous specimens carefully prepared mechanically. This enabled me to make out the cranial anatomy of a Devonian fish as far as that is possible on a fossil. This is of special importance, since *Eusthenopteron* belongs to the group of fish, the Osteolepiformes, from which not only the ichthyostegids but also the majority of tetrapods, the Eutetrapoda, is derived. The neural endocranium, the visceral endoskeleton with dental plates, and the subbranchial series were described by Jarvik

(1954). Moreover, the studies of the visceral arches and the dental plates associated with them and the neural endocranium opened the gates to a new field of research: the origin and composition of the gnathostome head. After studies of the embryological literature, in particular Holmgren's 'Studies on the vertebrate head' (1940, 1941, 1942, 1943), mainly new ideas were set out regarding the composition of the endocranium, the palatoquadrate and the parasphenoid in various groups of gnathostomes: placoderms, dipnoans, porolepiforms, coelacanthiforms, actinopterygians, brachiopterygians (Jarvik 1954; see also 1959b, 1960, 1964, 1972, 1980b).

The collection and preparation of the ichthyostegid material had been continued, and all essential parts of the postcranial skeleton (shoulder girdle, humerus, radius, ulna, pelvic girdle, hind limb, ribs), all well preserved, had been freed from matrix. It was then time to make a restoration of the skeleton of the whole animal. This restoration, complete except for the toes on the forelimb and carpal elements, was published in 1955 (Jarvik 1955a), and the same year I published a review of the ichthyostegalians in the Traité de Paléontologie (Jarvik 1955b).

In this year we finished our collection in the Greenland Upper Devonian, and to my disappointment we had failed to find a well-preserved ichthyostegid skull with lower jaws suitable for investigation by the Sollas Grinding Method. Preparation of many specimens had revealed that the ethmoid and orbitotemporal regions of the neural endocranium are too incompletely preserved to be described. Moreover, the otoccipital division and the overlying part of the exoskeletal cranial roof present structures without equivalents in either *Eusthenopteron* or post-Devonian tetrapods; these parts of the skull are, accordingly, difficult to interpret. It has therefore not been possible to present a restoration of the neural endocranium in *Ichthyostega*. The visceral endoskeleton is also imperfectly preserved.

However, in the excellent specimen No. P.222 of *Eusthenopteron*, the exo- and endoskeletal shoulder girdles and the pectoral fin are well preserved. Also the pelvic girdles and parts of the endoskeleton of the pelvic fin are present. The discovery (1948, 1949) of a well-preserved hindlimb and other parts of the appendicular skeleton in *Ichthyostega* made it possible to compare directly the limb skeleton in the earliest known tetrapods with the appendicular skeleton in the ancestral fishes. This prompted a discussion of the intricate and much debated dual problem of the origin of the paired fins and the pentadactyl tetrapod limbs. When I began to tackle these problems in the middle of the 1950s, it was generally agreed that, both in the paired fins of fish and in the tetrapod limbs, there is an important metapterygial stem or axis composed of a row of metameric elements, once situated in the body wall. For such an interpretation it was considered to be of fundamental importance to determine the position and extent of the presumed stem. However, after a perusal of the comprehensive relevant literature I found such a diver-

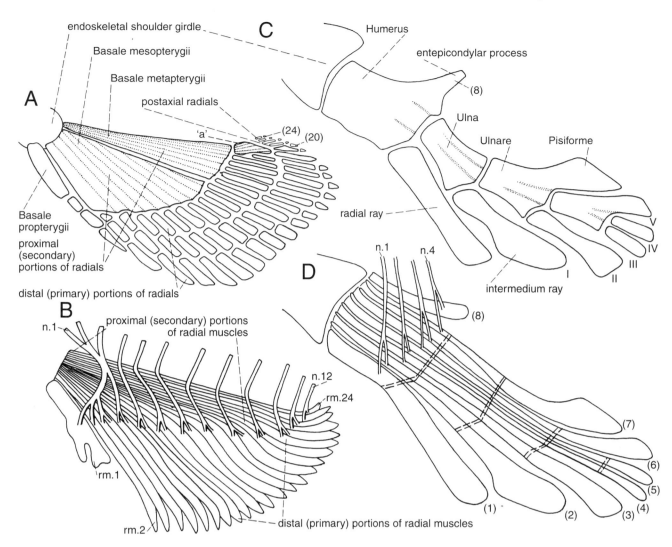

Fig. 12. Interpretations of the composition of the pectoral fins in *Squalus* (A, B) and *Eusthenopteron* (C, D). □A. Skeleton of the right pectoral fin. Assumed lines of fusion of proximal portions of radials indicated in dotted lines. □B. Diagrammatic representation of pterygial nerves and dorsal radial muscles of left pectoral fin. Compilation on the basis of Müller's (1911) figures and descriptions of the ontogenetic development and the author's own dissections. □C. Right pectoral fin in external view (cf. Jarvik 1964). □D. Diagrammatic representation to illustrate its composition. From Jarvik 1965b, Fig. 11. '*a*', complex skeletal element, composed of segments of adjoining radials (earlier regarded as a part of the metapterygial stem); *n.1, n.4, n.12*, metapterygial nerves; *rm.1, rm.2, rm.24*, radial muscles; *(1)–(8), (20), (24)*, radials or rays; *I–V*, digits.

sity of opinions as regards this stem, its position, and the number of metameric elements that enter into its composition, that I began to doubt that a metapterygial stem of great phylogenetic importance really exists. I also questioned the proposition that the so-called mesomeres (see Jarvik 1980b, p. 113, Figs. 63, 65) in the paired fins of *Eusthenopteron* and other fishes really are metameric. Intense studies of the literature on the ontogenetic development of the skeletal elements, nerves, muscles and blood vessels, and my own dissections and studies of variations in the paired fins, in particular in sharks, convinced me that the notion of a metapterygial stem of phylogenetic importance is due to misinterpretation of the fins in sharks. This led to a new, and in most respects quite different, solution of the problem (Fig.

12; see also p. 68). The skeletal elements of the paired fins and in the tetrapod limbs are composed of metameric rays or radials which, owing to a shortening of the joint between girdle and fin from behind, have been crowded together in their proximal parts. This led to a fusion of the proximal parts of the radials, but in different ways in the various groups, and most important, in a somewhat different fashion in the pelvic and pectoral fins of the same animal. It is also important to note that there are two rays (radials) in each metamere. The pectoral fin in *Eusthenopteron* is seven-rayed but if we also regard the entepicondylar process of the first mesomere (the humerus) as a ray, the fin will be composed of eight rays (Fig. 12D). Because there are two rays (radials) in each metamere, it was concluded that the fin belongs to four metameres and that the musculature is innervated by four spinal nerves, as in tetrapods.

Since the forelimb in tetrapods is derived from the pectoral and the hindlimb from the pelvic fin in the ancestral fish, and the pattern of the skeletal elements is somewhat different in the pelvic and pectoral fins, it was necessary to treat fore and hind limbs separately when discussing the origin and composition of the tetrapod limbs. Comparisons between the pelvic fin in *Eusthenopteron,* the hind limb in *Ichthyostega,* and the foot of the human embryo revealed a most remarkable agreement. It was also shown that the somewhat different pattern of the skeletal elements in the pectoral fin of *Eusthenopteron* in a most striking way is recapitulated in the embryonic development of the human hand. It then became easy to identify in the pectoral and pelvic fins of *Eusthenopteron* not only the most prominent elements (humerus, radius, ulna, femur, tibia, fibula) but also minor elements, such as the postminimus and the prehallux. Also, it was shown that the fingers and toes, which by previous writers (Holmgren, Westoll, and others) had been thought to be new formations (neomorphs), are represented in the ancestral fishes. The pelvic girdle in *Ichthyostega* was figured and was shown to be similar to that in *Eusthenopteron.* Moreover, it was shown that the humerus in *Eusthenopteron* is similar to that in the Permian stegocephalian *Eryops* and that only small modifications are necessary to transform the paired fins in osteolepiforms into pentadactyl limbs characteristic of *Ichthyostega* and other Eutetrapods. In urodeles the conditions are somewhat different, and they were discussed separately.

Some of the new results, as to the origin and composition of the paired fins and the tetrapod limb, were indicated in 1959 (Jarvik 1959b) but it was not until in the middle of the sixties (Jarvik 1964, 1965a, 1965b) that my new ideas were fully presented and documented.

One of the reasons for the delay was that in my studies of the fins I had been interested in the fin rays (lepidotrichia), and a paper on that subject was published (Jarvik 1959a). In that paper, a photograph (Fig. 19) of one of the tails of *Ichthyostega* figured in 1952 was included. Further preparation of that tail revealed that the distal ends of the dorsal radials are independent, a condition which implies that the agreement in the skeleton of the tail between *Ichthyostega* and *Eusthenopteron* (Fig. 38) is greater than I could document in 1952.

Another reason to the delay was that during my last summer in East Greenland (1956), we had found excellent material showing cranial structures of the Middle Devonian porolepiform *Glyptolepis groenlandica* (Jarvik 1972). Mechanical preparation exposed the urohyal and other structures, supporting my view that the porolepiforms are related to the urodeles. When I found that some of the skulls of *G. groenlandica* are well preserved and suited for investigation with the Sollas Grinding Method, I did not hesitate, and one of the skulls was picked out for that purpose. The grinding (performed by H.C. Bjerring) started in 1959 and succes-

sively revealed many previously unknown structural details. These data, together with information gained by mechanical preparation of other skulls, were discussed in a series of publications in the 1960s (Jarvik 1962, 1963, 1964). In 1972, I was able to present a comprehensive account on the porolepiform head, based also on other Middle Devonian forms and on material from the Lower Devonian in Spitsbergen (*Porolepis*) and from the ichthyostegid-bearing Upper Devonian in East Greenland (*Holoptychius*). In my view, this provided strong evidence that the porolepiforms are related to the urodeles and that the tetrapods are diphyletic in origin. By these studies and those of *Eusthenopteron* and other osteolepiforms, we had got a detailed and reliable knowledge of the structure of the head (as regards *Eusthenopteron* also the postcranial skeleton) in the two groups of early fish, the Osteolepiformes and the Porolepiformes, from which at least the majority of tetrapods (the Eutetrapoda and the Urodela) have been derived. A solid basis had been created for studies of the ichthyostegids, as Lauge Koch and Erik Stensiö required, and also for discussion on a comparative-anatomical basis of the relations to other groups of vertebrates. When my paper on the porolepiform head in the middle of 1971 was accepted for printing and my retirement was approaching, I decided to summarize our knowledge of the early (mainly Devonian) vertebrates and discuss their importance in various respects. This work, interrupted by papers on the saccus endolymphaticus (Jarvik 1975) and the acanthodian fishes, resulted in a comprehensive paper, finished in June 1979 and printed in an excellent way by the Academic Press in two-volume book late in 1980 (Jarvik 1980a, 1980b).

In that book a special chapter was devoted to the ichthyostegalians (Jarvik 1980a, pp. 218–244), with a new slightly modified restoration of *Ichthyostega,* photographs and restorations of pelvic girdle and hindlimb, figures of lower jaw, vertebral column, ribs, endo- and exoskeletal shoulder girdle, humerus, radius, ulna, femur, tibia, and fibula, all based on photographs of well-preserved and carefully prepared fossil material. One object of this present paper is to produce the photographs alluded to above, other photographs of well-preserved specimens illustrating, *inter alia,* the basipterygoid articulation and the remarkable structures encountered in the otoccipital division of the neural endocranium and associated dermal bones. For the sake of completeness, some photographs published earlier are also included.

The discovery of well-preserved material of tetrapods in the Upper Devonian quite naturally attracted great attention, and in particular *Ichthyostega* has been much debated in the relevant scientific literature. Schultze (1969) in his comprehensive account on labyrinthodont teeth presented sections of teeth in *Ichthyostega* and found similarities in particular with *Panderichthys.* The sections made by Schultze have been utilized also by Roček (1985), who on this and other material found similarities in the general pattern of tooth replacement between *Eusthenopteron* and *Ichthyostega*. Other papers to be mentioned are the stimulating discussions about the palaeoecology of *Ichthyostega* by Spjeldnæs (1982) and of *Acanthostega* by Bendix-Almgreen *et al.* 1988, 1990; and about electric tetrapods by Bjerring (1986).

If true ichthyostegalians have been found outside East Greenland is still uncertain. However, in 1972 Warren & Wakefield described three trackways from the Upper Devonian of Australia. In one of them there is a distinct undulating mark ('tail drag') between the left and right foot-prints. This tallies well with the fact that the tail-fin in *Ichthyostega* (Jarvik 1952, p. 16) is modified in such a way that it is obvious that the tail was dragged on the ground when the animal was walking (Fig. 47); and it cannot be excluded that the animal responsible for the foot-prints was an ichthyostegid. The trackway from the Early Devonian or Silurian in Australia recently described by Warren *et al.* (1986) was not found *in situ* and is less distinct; the imperfect lower jaw described by Campbell & Bell (1977) may not be a tetrapod (but see Ahlberg & Milner 1994). Considering the high degree of specialisation in *Ichthyostega* and the fact that in the Late Devonian it occurs together with another, in many respects different, tetrapod (*Acanthostega* Jarvik 1952, Clack 1988a, 1988b, 1994; Clack & Coates 1993) can only be taken to mean that the tetrapods arose at a much earlier date (Jarvik 1980a, p. 222; cf. Ahlberg 1991; Ahlberg & Milner 1994), and it is not unreasonable to imagine that they were in existence at the dawn of the Devonian or earlier. Be this as it may, the find of a single dubious four-toed print in the Late Devonian of Brazil (Leonardi 1983; see, however, the contrary interpretation by Roček & Rage 1994) and the discoveries of skeletal remains of a six-toed animal, *Tulerpeton curtum* (anthracosaur?), from the late Famennian of European Russia (Lebedev 1984, 1985) as well as of an Upper Devonian tetrapod in North America (Daeschler *et al.* 1994) demonstrate that the tetrapods were much diversified and had a world-wide distribution in Devonian times.

Description of the material of *Ichthyostega*

The material of *Ichthyostega* included in this paper belongs to the Geological Museum of Copenhagen University and will ultimately be housed there (MGUH VP collection). Currently, it is in the Paleozoological Department of the Swedish Museum of Natural History, Stockholm, and bears numbers A. 1 – A. 32, A. 34 – A. 84, A. 86 – A. 119, A. 122 – A. 132, A. 134 – A. 140, A. 142 – A. 143, A. 155 – A. 161, A. 163 – A. 169, A. 171 – A. 176, A. 178 – A. 221, A. 224 – A. 255 (note that the numbers 59 and 152 refer to the same specimen; both numbers have been used in the literature). The corresponding

Copenhagen numbers are MGUH VP 6001 to MGUH VP 6255 (with the same omissions). Several of the specimens consist of two or more separate parts; in the MGUH VP collection, the numbers of these specimens will be suffixed a, b, c, etc.

In addition to the material mentioned above, there are two unprefixed specimens, *viz.*, no. 99 and no. 220. These will bear the Copenhagen numbers MGUH VP 6262 and MGUH VP 6263, respectively.

The *Acanthostega gunnari* specimens collected prior to 1955 and considered briefly in this paper are A. 33 and A.85. They will bear the Copenhagen numbers MGUH VP 6264 and MGUH VP 6265, respectively.

Taxonomic remarks. – The two genera distinguished by Säve-Söderbergh in his preliminary account (1932a) were defined as follows (when different, the names of bones used in this paper are given in brackets):

Ichthyostega: 'Ichthyostegids with long, or at least not broad, skulls, with elongate, small to medium-sized orbits, situated with their greater parts in the posterior half of the skull. Each lachrymal (lacrimal) and jugal meeting farther forward than the anterior margin of the orbit. Each posterior antorbital (posterior tectal) with a suture towards the jugal. Fronto-parietals (frontals) probably only partially separated by a median suture, developed only in front of the pineal foramen. Internal nares wholly situated anterior to the anterior margin of the orbit'.

Ichthyostegopsis: 'Ichthyostegids with short and broad skulls, with large, rounded orbits, probably situated with their greater parts in the anterior half of the skull. Lachrymal (lacrimal) and jugal meeting laterally to the orbit. Posterior antorbital (posterior tectal) not meeting the jugal. Fronto-parietals (frontals) separated by a complete median suture. Internal nares ventral to the anterior parts of the orbit.'

The description of *Ichthyostegopsis* was based on a single specimen (A.5), the holotype of *I. wimani*, which is the type species. Four named species of *Ichthyostega* were described (*I. stensioei*, *I. watsoni*, *I. eigili* and *I.? kochi*), each including only one specimen (A.1, A.3, A.4, A.7).

The definition of the type species of *Ichthyostega*, *I. stensioei* (Fig. 10), reads: 'An *Ichthyostega* with a long skull and truncate snout. Dermal bones with fine external sculpture. Each anterior antorbital (anterior tectal) square, with a straight posterior margin and meeting a long and narrow anterolateral process of the posterior antorbital (posterior tectal), thus separating the naso-postrostral (anterior nasal) from the lachrymal (lacrimal). Posterior antorbitals (posterior tectals) with posterior and posteromedial orbital margins. Naso-postrostrals (anterior nasals) meet the frontals (posterior nasals) in sutures that have a direction from anteromedially towards posterolaterally. Jugals not excessively large.'

The three other species of *Ichthyostega* named by Säve-Söderbergh are distinguished from *I. stensöi*, and from each other, by more or less distinct differences in the shape of the skull and in the ornamentation and extent of the dermal bones of the skull table. Among the ichthyostegids that Säve-Söderbergh happened to pick up on or above the East Plateau of Celsius Bjerg (Fig. 1A), there are only five more complete skull remains. It may be considered as an odd coincidence that four of these examples should belong to four different species of *Ichthyostega* and the fifth to another genus *(Ichthyostegopsis)*. In the new material, I have been unable to identify with certainty any of the four species of *Ichthyostega*, and the presence of *Ichthyostegopsis* is questionable.

Since all species (also *I. wimani*) are defined mainly on differences in the skull table, it is of course impossible to decide if the examples of the postcranial skeleton, the detached lower jaws, and skulls showing mainly the palate belong to *Ichthyostegopsis* or to *Ichthyostega*; nor could they be referred to any of the species of the latter genus. In my paper on the fish-like tail (Jarvik 1952), I was therefore compelled to drop the specific names, and all the examples were described as *Ichthyostega* sp.

Later on, when dealing with disarticulated material of porolepiforms (*Porolepis* and *Holoptychius*) from Spitsbergen and East Greenland (Jarvik 1972) and osteolepiforms from East Greenland (Jarvik 1985), I further stressed the difficulties in distinguishing and defining genera and species on the basis of a large but incomplete and disarticulated group of fossils which, moreover, has been collected in many places within large areas. The introduction of new generic and specific names based on such a collection (Jarvik 1985, p. 6) 'would hardly serve any sensible purposes and would rather entangle the description'. As regards *Porolepis* and *Holoptychius* I further stated: 'It is the structure of the genera as a whole that is of interest. The variations are of course important but can be considered without the introduction of a new specific name for each of the more conspicuous morphological variants.'

Our catalogue of Devonian stegocephalians includes 261 items. Two specimens (A.33, A.85) have later (Jarvik 1952) been referred to *Acanthostega*, and the possibility that some other specimens belong to that genus cannot be ignored (as regards A.88 and A.90, see Clack 1988a, p. 713). Moreover, a few specimens have been identified as belonging to *Holoptychius* or *Eusthenodon* and some may be regarded as indeterminable. However, among the numbered items there are about 200 which contain one and in some cases two or more examples of indisputable ichthyostegids. Certainly, several variations have been encountered in this large and generally well-preserved material, but since some skulls have been compressed from the sides and others dorsoventrally, it is difficult to say to what extent the differences, e.g., in the breath of the palate or in the position of the jaw joint in relation to the posterior end of the braincase, are due to compression or other postmortem distortion. For reasons given above, it is meaningless to introduce new specific names on material, which, like that of *Holoptychius*, also

from the Remigolepis Series (Group), is disarticulated. Of the species of *Ichthyostega* it is appropriate to retain only the type species, *I. stensioei*. Moreover, the advisability of introducing a new genus (*Ichthyostegopsis*) on the basis of a single, incomplete skull is questioned.

According to the diagnosis, the skull in *Ichthyostegopsis* is 'short and broad'. However, Säve-Söderbergh's restorations (1932a, Figs. 4B, 6 and 11) show that in two of the species of *Ichthyostega* (*I. watsoni* and *I. eigili*) the skull is about as broad as long (length measured in the median line of the skull roof), and in this respect at least they do not differ from *Ichthyostegopsis*. Perhaps the skull in *Ichthyostegopsis*, as also pointed out by Säve-Söderbergh (1932a, p. 61), would be better characterized as short and high. Since in the new material there are a few skulls that seem to be just short and high, I was for some time inclined to believe that they belong to *Ichthyostegopsis*. Two of these skulls are figured in Pl. 1.

In one of these skulls, A.245 (Pl. 1:3), the lower jaws are preserved; this is of little help, as the jaws are unknown in *Ichthyostegopsis*. Moreover, the dermal bones have been broken into splinters in such a way that it is difficult to trace the sutures and determine the extent of the jugal and its relations to the posterior tectal. Since, furthermore, the fenestra exochoanalis is not exposed it is impossible to say if this specimen belongs to *Ichthyostegopsis*.

The other specimen, No.220 (Pl.1:1, 2), is of special interest because it was collected on the East Plateau by Kulling in 1929; accordingly, it is the first ichthyostegid skull found in East Greenland. (Kulling's material also included some incomplete ichthyostegid ribs; Pl. 41:1, 2). In this specimen, the skull table is lacking and the anterior part of the palate is presented in dorsal view. The fenestra exochoanalis is shown on both sides. A remarkable fact is, however, that the entopterygoids of both sides (as in A. 139; Pl. 24) meet in the median line at an angle of about 100°, which is in sharp contrast to the conditions in *Ichthyostegopsis* and most other ichthyostegids in which the palate is plane.

Because no other specimens in the material available can be referred to *Ichthyostegopsis*, and because this form in all essentials is a typical ichthyostegid and the differences to *Ichthyostega* are unclear, *Ichthyostegopsis* may not be a valid genus. However, the present evidence does not permit a definitive conclusion, and so I find it justified to retain both genera, each represented by only its type species. As in previous papers (Jarvik 1952, 1980a), all specimens that cannot be definitely assigned to either of these species are described as *Ichthyostega* sp.

The skull

General remarks

Skull remains are common in the new material. Unfortunately, however, there is no complete skull (with lower jaws) well enough preserved to be suited for investigation by the

Sollas Grinding Method. One of the best skulls is No. A.64 (Pls. 2:2; 3; 4) in which, however, the lower jaws are lacking. When this skull was collected in 1934, the main part could be easily removed from the bowl formed by the skull table, which is preserved in the counterpart.

It may seem strange that the skull table, not only in A.64 but also in A.158 (Pls. 14–18) and other specimens, may be removed from the underlying main part of the skull. Säve-Söderbergh (1932a) demonstrated that the dermal bones of the posterior parts of the skull roof are provided with strong descending laminae and a strong ventral process, which penetrate deep into the underlying part of the neural endocranium. The reason for this is that the tissues forming the inner parts of the dermal bones as well as the adjoining parts of the underlying endoskeleton are fragile in the present state of preservation and have often turned into powder. In contrast to conditions in *Eusthenopteron,* it is therefore sometimes difficult to distinguish between exo- and endoskeleton.

When this fragile endoskeletal bony tissue has been removed, either because of weathering or as a result of mechanical preparation, only a thin *bluish film* remains. This characteristic film, which probably reflects the periosteal lining, is found on the neural endocranium as well as on the palatoquadrate. However, it is fortunate that parts of these structures are formed by a more resistant bony tissue with a shiny external periosteal lining. When the skull table is removed, the underlying posterior part of the neural endocranium, that is, in the main in the otic region, presents a prominent paired, somewhat hemispherical elevation, *the otic elevation* (Pl. 15). The dorsal side of that elevation presents the broken dorsal ends not only of the paired ventral process described by Säve-Söderbergh but also of other ventral processes of the dermal bones of the skull roof. Owing to the fragility of the skeleton, it has been possible to remove these processes. As will be discussed below, this has resulted in deep holes in the otic elevation. Anorther remarkable fact is that the posterior part of the endocranium with the otic elevations is often separate. This separation is not, as could be imagined, due to the retention of the fissura preoticalis, and it has nothing to do with the intracranial joint in the osteolepiform ancestors. It is a consequence of the fragility of the bony tissue, the short parasphenoid, and the fact that the palatoquadrate, behind the basal articulation, is bent laterally to form an almost transverse wall immediately in front of the otic elevation.

The skull table, with discussion of the terminology of the dermal bones

As established by Säve-Söderbergh, the ichthyostegid dermal bones, which in the osteolepiforms form the external cheek plate, have fused with the dermal bones of the skull roof forming a solid unit, the skull table.

The material available in 1931 permitted Säve-Söderbergh to make out the pattern of the dermal bones of the skull table,

and he could also describe essential parts of the sensory-canal system (Fig. 11). In his paper on *Osteolepis* (Säve-Söderbergh 1933a), the dermal bones and the sensory canals of the cheek in an osteolepiform were correctly described for the first time. In ichthyostegids he found exactly the same bones (maxilla, lacrimal, jugal, postorbital, squamosal, preopercular and quadratojugal) and the same course of the sensory canals as in *Osteolepis.* The osteolepiform terminology could therefore be directly applied to the ichthyostegids.

However, the dermal bones of the ichthyostegid skull roof were not so easy to interpret. In osteolepiforms, the pineal foramen is situated between bones which at that time had been called 'frontals', whereas in early fossil tetrapods they lie between the bones called 'parietals'. In order to explain this difference Säve-Söderbergh (1932a, Fig. 16) propounded a new theory (Fig. 13). Since he had found that the paired parietal in one specimen of *Eusthenopteron* (1932a, Fig. 19) is divided into an anterior and a posterior parietal, and with reference to the conditions in dipnoans, he claimed that also the frontal may be subdivided in a similar way, and that the pineal foramen is situated between the bones in the posterior pair of frontals. He further assumed that the posterior frontals (with the pineal foramen) had fused with the anterior parietals. Consequently, the pair of bones which enclose the pineal foramen in tetrapods, previously called 'parietals', are compound frontoparietals. In a later paper Säve-Söderbergh (1935a), using the same terminology, figured and analysed the patterns of the dermal bones of the head in a great number of labyrinthodont stegocephalians and primitive Reptilia and, moreover, he presented (p. 202) 'a totally changed classification of the Gnathostome Vertebrates'.

In a pungent criticism of the latter paper, Romer (1936b, p. 536) concluded that Säve-Söderbergh's 'phylogenetic conclusions are for the most part open to serious doubt, his scheme of classification impractical, and his bone terminology especially unfortunate'. With this started a heated debate on the homologies of the frontal and parietal bones, which is still going on (Borgen 1983; Schultze & Arsenault 1985, 1987; Klembara 1992, 1993; Bjerring 1995; and others).

In connection with his studies of the cranial roof in dipnoans, Romer (1936a, p. 253) had pointed out (1941, p. 158) 'that the fish extra-scapulars (customarily homologized with the tetrapod postparietals and tabulars) were not properly part of the skull pattern'. In a series of publications, Romer (e.g., 1941, 1946, 1956a, 1962) has expressed deep regret that he had not understood the implications of this condition and that it was (1941, p. 158) 'a brilliant suggestion of Westoll ('36, p. 166; '40, pp. 71–73)' that solved the problems (Fig. 14). If, as suggested by Westoll, the extrascapulars have been lost in tetrapods, it follows (Romer 1941, p. 158) 'that there had been no shifting of the pineal opening or mysterious 'reshuffling' of elements; that the supposed parietals were actually the tetrapod postparietals; and the supposed frontals were actually the parietals. In the shift from fish to amphibian the postparietals, originally long, have become much

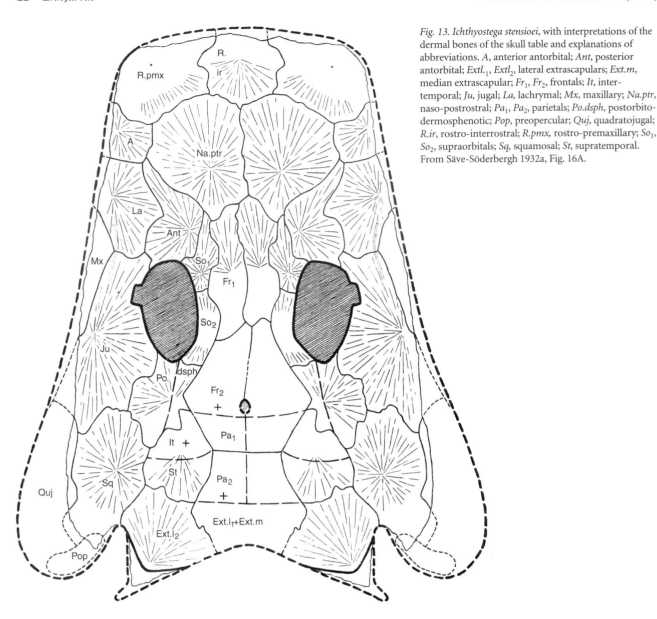

Fig. 13. Ichthyostega stensioei, with interpretations of the dermal bones of the skull table and explanations of abbreviations. *A*, anterior antorbital; *Ant*, posterior antorbital; *Extl.$_1$*, *Extl$_2$*, lateral extrascapulars; *Ext.m*, median extrascapular; *Fr$_1$*, *Fr$_2$*, frontals; *It*, intertemporal; *Ju*, jugal; *La*, lachrymal; *Mx*, maxillary; *Na.ptr*, naso-postrostral; *Pa$_1$*, *Pa$_2$*, parietals; *Po.dsph*, postorbito-dermosphenotic; *Pop*, preopercular; *Quj*, quadratojugal; *R.ir*, rostro-interrostral; *R.pmx*, rostro-premaxillary; *So$_1$*, *So$_2$*, supraorbitals; *Sq*, squamosal; *St*, supratemporal. From Säve-Söderbergh 1932a, Fig. 16A.

shortened (relatively at least); the parietals have in consequence become relatively more posteriorly placed, and the anterior part of the skull has become greatly elongated'.

As argued in previous papers (Jarvik 1967, 1980b) and as will be further elucidated below, the Westoll–Romer theory is fundamentally false. With regard to the differences in the position of the pineal foramen, which puzzled Säve-Söderbergh as well as Westoll and Romer, a simple solution has been presented (Jarvik 1967, pp. 197, 199, 205, Figs. 10, 13, Pl.4; 1980b, p. 206, Fig. 124). When the brain stem is straight, as in osteolepiforms, anurans and other batrachomorphs, the pineal foramen is found between the frontals. In the group of tetrapods that Säve-Söderbergh called 'reptilomorphs', in contrast, the brain with the pineal complex has been pushed backwards within the braincase. This has caused a distinct flexure of the brain stem, and as a conse-

quence the pineal foramen in them is generally situated between the parietals.

For the interpretation of the dermal bones of the skull roof, we cannot rely on their relations to structures situated far below, such as the basipterygoid process or the fossa hypophysialis (Fig. 14). We must consider their position in relation to the directly underlying structure, that is the endoskeletal skull roof. An important, previously overlooked fact is that this roof is a complex formation, composed of a series of cranial tecta (Jarvik 1980b, Figs. 59–61). I have demonstrated (Jarvik 1980b) that the various dermal bones can be defined according to their relations to the directly underlying cranial tecta. Because the parietal in mammals arises in relation to the two synotic tecta and the tectum transversum, it was easy to see that the bone in *Eusthenopteron* and other osteolepiforms that I have always termed *parietal* has a corre-

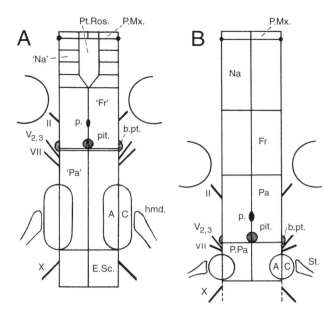

Fig. 14. Diagrams showing relationship of dermal skull-roof bones to important endocranial structures, emergences of cranial nerves, etc., in (A) an osteolepid and (B) a stegocephalian. *A.C*, auditory capsule; *E.Sc*, extrascapular; *Fr*, frontal; *Na*, nasal; *Pa*, parietal; *P.Mx*, premaxillary; *P.Pa*, postparietal; *Pt.Ros*, postrostrals; *St*, stapes or columella; *b.pt*, basipterygoid process or articulation; *hmd*, hyomandibula; *p*, pineal foramen; *pit*, position of pituitary; II, V2, 3, VII, X, cranial nerves. From Westoll 1943, Fig. 3.

sponding position and, accordingly, is homologous with the parietal in mammals (see, however, Bjerring 1995). As is well known, the parietal in *Eusthenopteron* and other osteolepiforms (Säve-Söderbergh 1932a, Fig. 19; 1933a; 1941; Jarvik 1948a, pp. 63–64) is often subdivided into anterior and posterior parietals. It is then of great interest that the parietal in *Homo* (de Beer 1937, p. 503; Jarvik 1967, Pl.2E; 1980b, Fig. 125C) arises from two primordia, upper (or anterior) and lower (or posterior). The dermal bone following behind the parietal is in reptiles generally called 'postparietal' whereas in mammals it is called 'interparietal'. This element is particularly well known in man. It arises in connection with the most posterior of the cranial tecta, the tectum posterius, and fuses with parts of that tectum forming a mixed bone, the supraoccipital. Starck (1975, Fig. 542) has figured this element in a 100 mm embryo of *Homo* (Fig. 15). The dermal (interparietal) part of the mixed bone includes a median element associated with the tectum posterius and on each side of that a large lateral element known as the preinterparietal (see also de Beer 1937, p. 444).

Behind the parietal in *Eusthenopteron* lies the extrascapular series, including a median extrascapular and paired lateral extrascapulars. As is well shown on the serial sections (Jarvik 1975, Fig. 13C, D), the median extrascapular rests on the posteromedian part of the endoskeletal skull roof including the supraoccipital plug, which are median parts of the tec-

tum posterius (Jarvik 1975, pp. 203–204; 1980b, pp. 100–101, 268, Fig. 61). Consequently, the interparietal series in man has exactly the same position in relation to the neural endocranium as the extrascapular series in *Eusthenopteron*, and like the latter it includes a median element suggestive of the median extrascapular and a paired lateral element, the preinterparietal, obviously corresponding to the lateral extrascapular of our piscine ancestors.

This most remarkable similarity between the Devonian fish *Eusthenopteron* and an advanced mammal such as man may seem surprising to students who like Schultze & Arsenault (1985, p. 295) argue 'for step-by-step comparison of closely related forms'. However, this striking similarity also tallies well with the close agreement between *Eusthenopteron* and man in many other respects (Jarvik 1980b, pp. 267–269). Since the extrascapular series (the interparietal or postparietal) is thus retained in man and in many other tetrapods (p. 24; de Beer 1937, p. 444) Westoll's suggestion that the extrascapular series has been lost in tetrapods is untenable, and so are the interpretations of the dermal bones of the skull roof, that are based on his idea.

In addition to the misinterpretations of the dermal bones of the skull roof, the Westoll–Romer theory embodies several other unacceptable claims. Thus, when Romer (1941, pp. 158–159) stated that the 'postparietals' in the tetrapod phylogeny have been successively reduced to disappear by the end of the Triassic, he overlooked the fact that the postparietal (interparietal) is retained in many extant tetrapods and is a prominent element in the human skull (Fig. 15). The main reason for this oversight is that the long parietals in osteolepiforms have erroneously been called 'postparietals'. An essential part of the Westoll–Romer theory is that the shortening of the alleged postparietal is combined with a shortening of the otic region with the auditory capsules and the membraneous labyrinth (Fig. 14). These processes have been illustrated by Westoll (1943, Figs. 3, 4), but, as is readily seen, they would lead to a complete obliteration of the auditory capsules and the membraneous labyrinths if, as is claimed, the postparietals migrate onto the occipital surface of the skull and finally disappear. In my view, this is quite absurd. As demonstrated (Jarvik 1967, Fig. 9; 1980b, Fig. 126), there are no remarkable differences in the length of the otic region between osteolepiforms (*Eusthenopteron*), anurans (*Rana*), lizards (*Lacerta*) and mammals (*Lepus* or *Didelphys*).

Another idea embodied in the Westoll–Romer theory is that one of the most posterior nasals in osteolepiforms has suddenly increased in size and migrated backwards from the ethmoidal to the orbitotemporal region forming the frontal. In consequence, the osteolepiform frontals with the pineal foramen have been shuffled backwards to the otic region to form the parietals. This had the effect that the osteolepiform parietals, named postparietals, were 'pushed down onto the occipital surface to disappear by the end of the Triassic' (Romer 1941, p. 159).

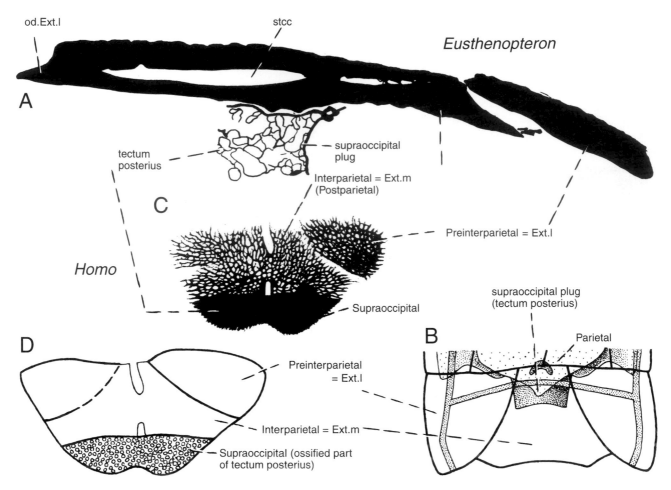

Fig. 15. Diagrammatic representations illustrating the remarkable fact that the osteolepiform extrascapular series is retained in the human embryo and takes part in the formation of the occipital protuberance. □A, B. *Eusthenopteron foordi*. A, transverse section through the posterior part of the neural endocranium and adjoining dermal skeletal elements demonstrating that the median extrascapular is supported by the supraoccipital plug, a part of the tectum posterius. Section No 347 of grinding series 2 (P. 222; see also Fig. 17A). B, posterior part of neural endocranium with extrascapular series. Dorsal view. □C. Photograph of transparent occipital protuberance of human embryo (90 mm). Inner view. From Starck 1975, Fig. 542c. □D. Drawing after C. Note that only the median element of the extrascapular series in man as well as in *Eusthenopteron* is supported by the neural endocranium; note also that the medial margin of the lateral element in both man and *Eusthenopteron* is convex (this in contrast to conditions in porolepiforms; Jarvik 1980a, Fig. 185). *Ext.l, Ext.m,* lateral and median extrascapulars; *od.Ext.l,* area overlapped by lateral extrascapular; *stcc,* supratemporal commissural canal.

This story, which basically is a consequence of the erroneous exclusion of the extrascapular series, is also untenable, as I had begun to suspect in 1942.

As is well known, the fronto-parietal in anurans is a compound bone, arising from independent frontal and parietal primordia, (for references, see Jarvik 1942, pp. 343–344, footnote; 1967, p. 144). In my paper on the snout, I also discussed the dermal bones (Jarvik 1942, pp. 343–344, 550), and after having demonstrated that the anuran snout is of the osteolepiform type, it could be established that the frontal in *Eusthenopteron* must be homologous with the frontal component of the anuran fronto-parietal. The frontal in *Eusthenopteron* is followed posteriorly by the three paired bones (intertemporal, supratemporal and parietal) that form the parietal shield, and behind that shield by the extrascapular series. If we turn to anurans and consider all the dermal

bones that have been described (Jarvik 1967, pp. 194–197; 1968, p. 508, Fig. 2; 1972, pp. 148–149, Fig. 104B; 1980b, Fig. 126), we find exactly the same bones (intertemporal, supratemporal and parietal) and in exactly the same position in relation to the directly underlying structures, as in *Eusthenopteron* (Fig. 17). Moreover, there is sometimes (*Pelobates*, Fig. 16A) a median extrascapular, as in *Eusthenopteron*, *Alligator* (Bellairs & Kamal 1981, p. 244; see also Mook 1921) and *Homo*, related to the tectum posterius. Also, in *Pelobates* there may be a rather large posterior median postrostral (Fig. 16B) situated as that in *Eusthenopteron*. As is thus quite evident, there has been no mysterious shuffling of the dermal bones of the skull roof either forwards or backwards (cf. Schultze & Arsenault 1985, pp. 294–295). The bones in the Eutetrapoda (as regards the Urodelomorpha, see Jarvik 1972, pp. 263–266) remain in the same position as in the

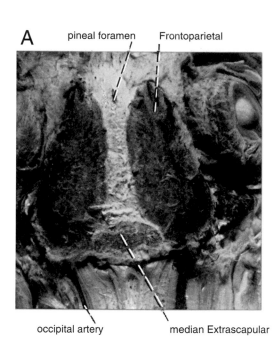

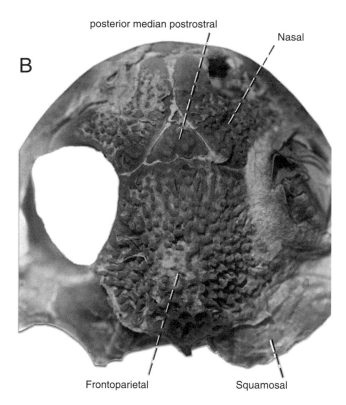

Fig. 16. □A. *Pelobates syraicus.* Photograph of part of head showing ornamented median extrascapular. From Jarvik 1967 (Pl. III A). □B. *Pelobates fuscus.* Photograph of head with independent posterior median postrostral.

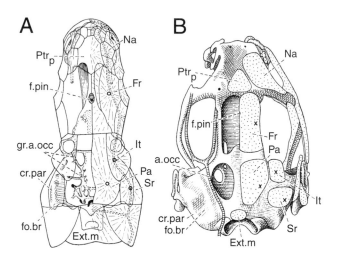

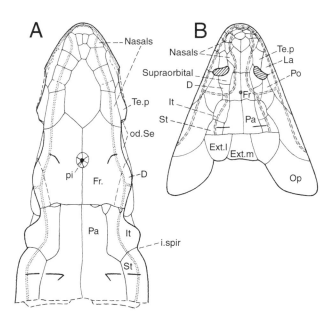

Fig. 17. Representations to demonstrate the similarities in the number, position and relations to the underlying part of the neural endocranium of the dermal bones of the skull roof between (A) Devonian osteolepiforms (*Eusthenopteron foordi*) and (B) Recent anurans (composite). From Jarvik 1968, Fig. 2. The areas of origin of the dermal bones in A are indicated by the position of the centre of radiation and in B marked with x. *Ext.m*, median extrascapular; *Fr*, frontal; *It*, intertemporal; *Na*, nasal; *Pa*, parietal; *Ptr*p, posterior median postrostal; *St*, supratemporal; *a.occ*, occipital artery; *cr.par*, crista parotica; *fo.br*, fossa bridgei; *f.pin*, pineal foramen; *gr.a.occ*, groove for occipital artery.

Fig. 18. The skull roof in two osteolepiforms, (A) *Eusthenopteron saevesoederberghi* (from Jarvik 1944, Fig. 19) and (B) *Panderichthys rhombolepis* (from Vorobyeva 1977, Fig. 2B; cf. Young *et al.*, 1992, Fig. 46), in which the nasal series extends back to meet the supraorbital in a short suture. *D*, dermosphenotic; *Ext.l*, *Ext.m*, lateral and median extrascapulars; *Fr*, frontal; *It*, intertemporal; *La*, lacrimal; *Op*, opercular; *Pa*, parietal; *Po*, postorbital; *St*, supratemporal; *Te.p*, posterior tectal; *i.spir*, spiracular notch; *od.So*, area overlapped by supraorbital; *pi*, pineal foramen.

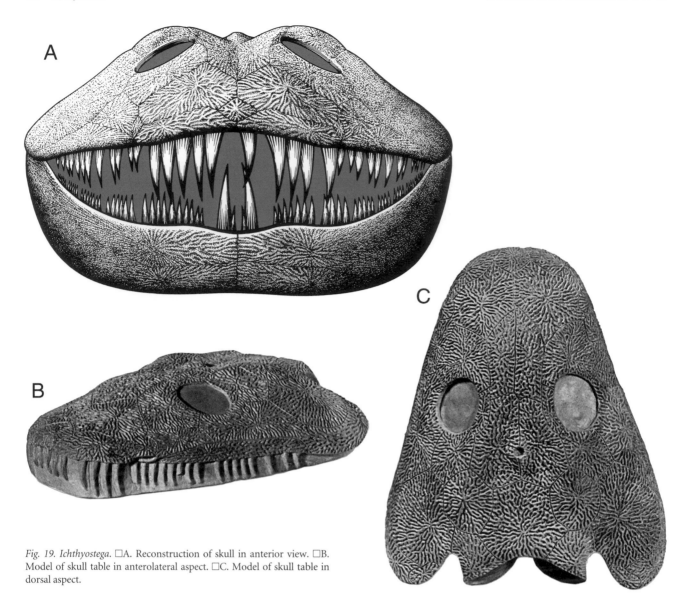

Fig. 19. Ichthyostega. □A. Reconstruction of skull in anterior view. □B. Model of skull table in anterolateral aspect. □C. Model of skull table in dorsal aspect.

osteolepiforms. As claimed already by Gaupp (1905, p. 617), the frontal thus arises in relation to the orbitotemporal region, whereas the parietal covers the otic region; moreover, the extrascapular series, as we have seen, is related to the tectum posterius. Of course there are many variations in the proportions of the skull and the extent of the individual bones, but the alleged shortening of the otic region (Fig. 14) has not occurred.

One variation of importance for the interpretation of the ichthyostegids may be mentioned, however. The nasal series belongs in the main to the ethmoidal region, and in *Eusthenopteron foordi* (Fig. 17A) only about the posterior half of the hindmost nasal (Na7 or Na 6+7; Jarvik 1944, 1980a, Figs. 116, 120) is situated behind that region. However, in *E. saevesoederberghi* (Fig. 18A) and some other osteolepiforms (*Osteolepis, Latvius, Glyptopomus, Megalichthys;* Jarvik 1967, pp.

201–203), the posterior nasal extends much farther backwards, and as in *Panderichthys* (Fig. 18B) and the closely related form *Elpistostege* this element – by Westoll (1938, 1943), Vorobyeva (1977) and Schultze & Arsenault (1985) erroneously called 'frontal' – is joined to the anterior part of the supraorbital by a short suture.

As said above, the Westoll–Romer theory rests on a false basis; it is therefore unacceptable and must be rejected. As regards Säve-Söderbergh's views, on the other hand, it is clear that he was correct when postulating that the extrascapular series is retained in tetrapods (Figs. 11, 13). However, before passing to the ichthyostegids, it is necessary to recall an important condition unknown before the middle of the 1940s.

As first discovered by studies of the dermal bones and sensory lines in the brachiopterygian *Polypterus,* and further

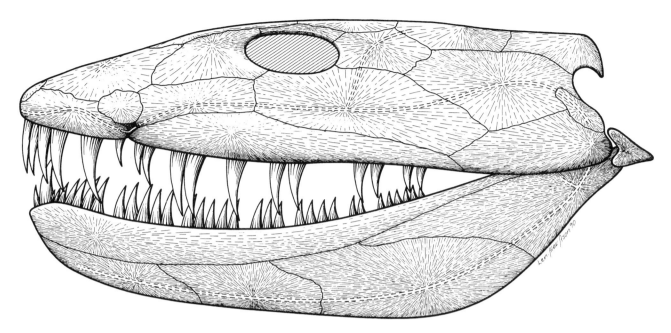

Fig. 20. Restorations of skull of *Ichthyostega* in lateral aspect. Ornamentation omitted. For explanations see Figs. 21 and 31A.

documented by the remarkable variations in the pattern of the dermal bones and the course of the sensory lines in skulls of *Acipenser sturio* (Jarvik 1948a, pp. 68–73), the sensory lines (both canals and pit-lines) have in a remarkable way been deflected in connection with fusions of bones. Later (Jarvik 1967, Figs. 5–7, Pl. 3F; 1972, pp. 147–150; 1980a, pp. 250–253, 406, 414–415; 1980b, p. 46), such deflections have been demonstrated in other actinopterygians, as well as in dipnoans, porolepiforms and osteolepiforms, which enabled me to confirm Säve-Söderbergh's view that the three posterior bones of the skull roof in *Ichthyostega* include extrascapular components (Jarvik 1967, p. 191, Fig. 6).

Dermal bones, sensory canals and openings of the skull table

The dermal bones of the skull table are ornamented with anastomosing ridges and isolated tubercles (Fig. 19, Pls. 2, 3, 5, 6, 8, 11, 13:1). Pores of the sensory canals appear in certain places, in particular on the cheek, but traces of pit-lines are discernible only on the lower jaw (pp. 47–48, Fig. 32, Pls. 32:1; 33; cf. Klembara 1992). An isolated lateral rostral is present in one specimen (Pl.10). As in osteolepiforms, the supraorbital is often missing and the preopercular is rarely preserved. However, on the whole, the dermal bones of the skull table are intimately connected and form a solid unit including equivalents to the frontoethmoidal and parietal shields, the extrascapular series and the external cheekplate in osteolepiforms. As in other stegocephalians, the skull table is pierced by several openings (pineal foramen, p. 31; paired orbital fenestra, p. 31; and paired fenestra exonarina, p. 43). Because of its ventral position, the latter is not visible when

the skull is viewed from above. A well-developed otic notch is present (Pl.13:1, 2). Behind the skull table, a vestigial subopercular has been found in one specimen (present on both sides, Pl.13:3–5).

The ornamentation provides some information about the bone pattern, but in several specimens it is difficult to make out safely the position and course of the sutures between the individual bones. However, in some specimens (Pls. 6; 8; 11:3) the sutures are distinct, and it can be confirmed that Säve-Söderbergh (1932a), on the basis of the material available in 1931, discovered and identified all the dermal bones that enter into the composition of the rigid skull table. As pointed out by him, there are certainly some variations in the shape and extent of the bones. However, since it has been found meaningless to distinguish separate species (p. 20), and since, moreover, it is in many cases difficult to ascertain the exact position and course of the sutures, these variations will be dealt with only in passing.

The restorations of the skull table which – partly based on the new material – were presented in previous papers (Jarvik 1952, Fig. 35; 1980a, Fig. 171) and which, with a few modifications, will be used in the present paper (Figs. 20, 21), agree in all essentials with those given by Säve-Söderbergh (1932a, Figs. 2, 4, 6, 7, 15; 1934a, Fig. 1D). The terminology of the dermal bones that was introduced in a special paper (Jarvik 1967) is retained (for discussion see pp. 23–26).

Sensory canals

As demonstrated by Säve-Söderbergh (1932a, p. 83) 'at least the greater part of the lateral line system of the head in the ichthyostegids was situated in closed canals in the dermal

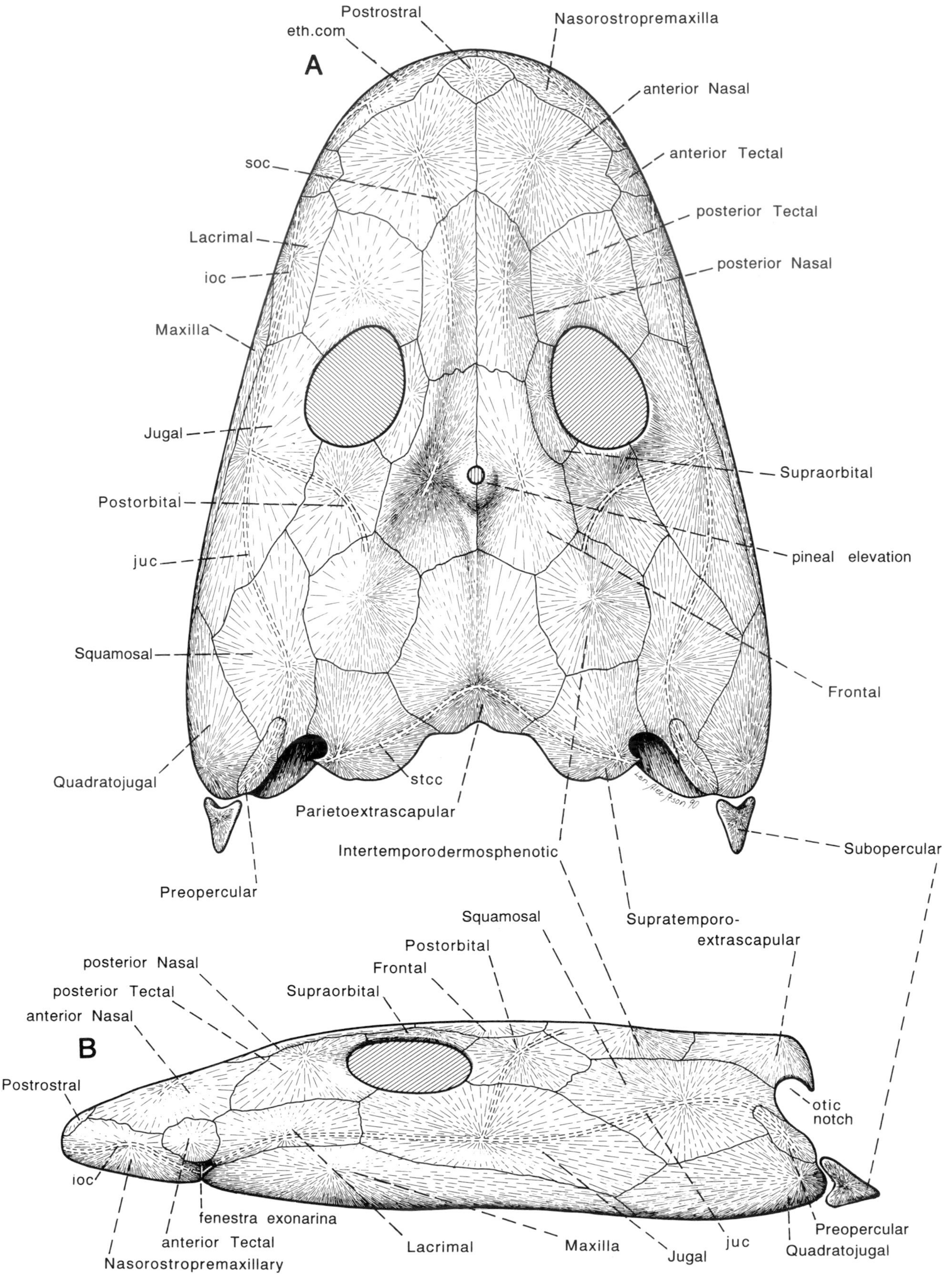

Fig. 21. The skull table of *Ichthyostega* in dorsal and lateral aspects. *eth.com*, ethmoidal commissure; *ioc*, infraorbital sensory canal; *juc*, jugal sensory canal; *soc*, supraorbital sensory canal; *stcc*, supratemporal commissural canal.

bones'. These canals, which housed the membranous sensory canals with their neuromasts (cf. Jarvik 1980a, p. 75, Figs. 8–11), are filled with a fragile matrix in the fossils. After skilled preparatory work, Säve-Söderbergh exposed the fragile filling of the canals (1932a, Pls. 19, 21) and was able to describe essential parts of the sensory canals with their tubes and superficial pores (Fig. 11). However, after the publication of his preliminary account (1932a) and before his illness, Säve-Söderbergh succeeded in disclosing the main part of the supratemporal commissure (*stcc*) in one specimen (A.55, Pl.6:3) collected in 1934 in the prolific locality at 1174 m in Sederholm Bjerg. In this and another specimen (A.57) from the same site, he uncovered the anterior part of the infraorbital canal, including portions of the ethmoidal commissure (Pls. 6:2; 9:1, 2). A third specimen (A.63), also collected in 1934, shows parts of the sensory canals of the cheek (Pl.9:3). A remarkable fact displayed by this specimen is that the infraorbital canal passes upwards through the postorbital to the dorsal margin of that bone. However, all attempts to find out if that canal joins the supraorbital canal, as it does in osteolepiforms, have failed. In osteolepiforms, this junction occurs in the dermosphenotic, and the fate of this bone in ichthyostegids is therefore still unknown (cf. Borgen 1983). The dermosphenotic may be incorporated in the postorbital, as suggested by Säve-Söderbergh, or it may have fused with the frontal. However, these interpretations are contradicted by the conditions in two other specimens. In A.7 (*Ichthyostega kochi*), the infraorbital canal (Fig. 23A; Säve-Söderbergh 1932a, Fig. 10, Pl. 21:2) curves backwards in the postorbital towards the bone that Säve-Söderbergh termed 'supratemporo-intertemporal'. In the second specimen (A.199, Pl. 5:4), the course of the canal is indicated by a groove in the postorbital with the same position and exactly the same backwards curvature as the canal in A.7. Of interest is that the groove continues onto the bone that Jarvik (1967, Fig. 11) termed 'intertemporo-dermosphenotic' (cf. *Peltostega*, Fig. 22E, *s.ot*). This interpretation is supported by the fact that the two components in *Sclerocephalus* (Boy 1988, Fig. 2H) are separated.

Although studies of the material collected after 1931 have added certain important new data, it has to be admitted that our knowledge of the sensory-canal system of the skull table in ichthyostegids is still incomplete. This is true in particular of the cranial roof, and a remarkable fact is that generally no sensory-canal pores are discernible in the ornamentation on this side of the skull. A possibility is, of course, that the sensory-canal system was incompletely developed in the posterior part of the skull roof, as is generally the case in post-Devonian stegocephalians (see e.g., *Peltostega*, Fig. 22E).

The dermal bones of the skull roof

Säve-Söderbergh's view (1932a) that the premaxilla has fused with rostral components to form a compound rostro-premaxilla has been confirmed (cf. Jarvik 1980a). It it also includes a nasal component, as he suggested later (Säve-Söderbergh 1934a, Fig. 1D; 1935a, p. 11), is difficult to say (cf. below). As regards the anterior median element, which Säve-Söderbergh termed 'rostro-interrostral', the ethmoidal commissure, judging from A.55 (Pl. 6:2), gives off tubes into this bone but does not traverse it. Accordingly, the proper name is *postrostral*, and it corresponds to the anterior median postrostral in *Eusthenopteron* and to the internasal in post-Devonian labyrinthodonts (Säve-Söderbergh 1935a, p. 23). A remarkable fact is, however, that this bone is paired in *Acanthostega* (Clack 1988b, Fig. 7), as in loxommatids (Beaumont 1977). Regarding the dermal bones at the tip of the snout, *Ichthyostega* thus agrees strikingly with *Eusthenopteron*. In both, there is a median postrostral between the tooth-bearing bones, and since the latter in *Eusthenopteron* are compound naso-rostro-premaxillae (Jarvik 1942, p. 347; 1944, p. 7, Fig. 3B), it is reasonable to use this name also in *Ichthyostega*. Behind the three bones at the tip of the snout lies the large paired bone which Säve-Söderbergh termed 'naso-postrostral'. Middle and posterior postrostrals are well developed in *Eusthenopteron* and other osteolepiforms (Fig. 18; Westoll 1936; Jarvik 1944, 1948a). A posterior median postrostral ('interfrontal') is present in some post-Devonian stegocephalians (Säve-Söderbergh 1935a, p. 42; Romer 1947, p. 21; Welles & Cosgriff 1965, Fig. 25) and certain anurans (Fig. 16B; Jarvik 1967, p. 194; 1968, p. 508; Špinar 1972, pp. 193, 202). In ichthyostegids (Fig. 20) no such element has been found, and for the sake of simplicity the possible postrostral components have been disregarded and the bone has been named *anterior nasal*. The paired bone next behind ('anterior frontal' of Säve-Söderbergh 1932a; 'frontal' of Westoll 1943 and Romer 1947, 1966) extends backwards to join the supraorbital in a short suture; that is, it extends exactly as far backwards as the nasal series in *Eusthenopteron saeve-soederberghi*, *Panderichthys* (Fig. 18) and several other osteolepiforms. For reasons given previously (Jarvik 1967, pp. 201–203), it is to be regarded as a *posterior nasal*. The supraorbital resembles that in *Eusthenopteron* and other osteolepiforms. The bone in front of the supraorbital ('prefrontal' of Westoll and Romer) was called 'supraorbito-antorbital' by Säve-Söderbergh, but since the name *antorbital* is to be replaced by the name *tectal* (Jarvik 1942, pp. 249, 351), it has been called 'supraorbito-tectal', as has also the similar bone in osteolepiforms (Jarvik 1967, Fig. 11).

The view that the bone is formed by fusion of an anterior supraorbital ('supraorbital 1') and a posterior tectal rests on Säve-Söderbergh's statement (1932a, pp. 16, 76) that the two bones are independent in one specimen (A.1) of *Ichthyostega* (Fig. 10). However, in all other specimens studied by Säve-Söderbergh and in numerous specimens in the new material showing this area, there is always a single bone without any traces of subdivision. Therefore, I have found it reasonable to use the name *posterior tectal* for this bone in ichthyostegids as well as in osteolepiforms (Jarvik 1980a, p. 159, Figs. 115, 116, 171). The *anterior tectal* ('anterior antorbital, ' Säve-Söder-

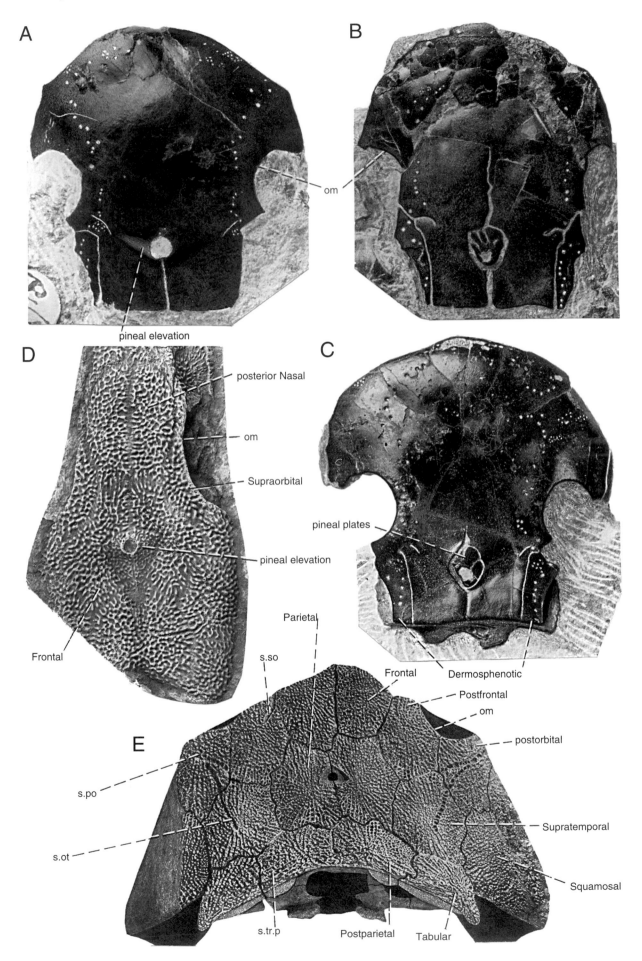

bergh 1932a; 'septomaxillary', Romer 1966) in ichthyostegids is a somewhat square bone, relatively shorter than the corresponding bone in osteolepiforms. In the latter it forms the dorsal margin of the fenestra exonarina, whereas in ichthyostegids it lies above a gap (the exonarial gap, p. 43) in the skull table between the naso-rostro-premaxilla and the ornamented part of the maxilla, the inturned anterior part of which is visible in the gap (Pls. 4:4; 9:2; 11:1; 17:2, 3). Säve-Söderbergh assumed that the infraorbital canal must have traversed this gap ventral to the external nostril (Fig. 30). In one specimen a sensory-canal bone has been found in the gap in almost the same position as the sensory canal as restored by Säve-Söderbergh. This bone has been termed the 'lateral rostral' although it only represents the sensory-canal portion of the osteolepiform lateral rostral. It will be described below with the dermal bones of the palate.

As demonstrated elsewhere (Jarvik 1967, and discussed above), the two bones lying behind the posterior nasals and surrounding the pineal foramen (see next column) represent the *frontals* ('fronto-parietals' of Säve-Söderbergh; 'parietals' of Westoll and Romer). The large median element, characteristic of ichthyostegids, which follows behind the frontal is pierced by the supratemporal commissure (p. 29) and is therefore a compound *parieto-extrascapular*, as claimed by Säve-Söderbergh (1932a, p. 78).

Lateral to the parietal in osteolepiforms lie two sensory-canal bones, the intertemporal and the supratemporal (cf. Long 1985b). As discussed above (p. 29), the intertemporal in ichthyostegids most likely has fused with the dermosphenotic into a compound *intertemporo-dermosphenotic* (Fig. 21). Provided this is true, it follows that the sensory-canal bone next behind must be a compound *supratemporo-extrascapular* ('lateral extrascapular', Säve-Söderbergh 1932a; 'tabular', Westoll 1943, Fig. 2; and others). The free posterior part of that compound bone forms the dorsal margin of the *otic notch*, which most likely, judging from the configuration of that margin (Pl. 13:1, 2), contained a tympanic membrane (cf. Godfrey *et al.* 1987).

The dermal bones of the cheek

As in *Osteolepis*, *Eusthenopteron* and most other well-known osteolepiforms (Säve-Söderbergh 1933a; Jarvik 1944, 1948a, 1950a, b, 1952, 1985; Jessen 1966; Vorobyeva 1977; Young *et al.* 1992, and others), the cheek in *Ichthyostega* (Fig. 21B) is covered with seven bones (lacrimal, maxilla, jugal, post-

orbital, squamosal, preopercular and quadratojugal), as a whole situated as in osteolepiforms (Fig. 24) and with the same relations to the sensory canals. As we have seen (p. 29), the infraorbital canal at least in some specimens of *Ichthyostega* curves backwards in the postorbital, as in *Peltostega* (Fig. 22E) and many other post-Devonian stegocephalians. Moreover, the infraorbital canal, also in contrast to conditions in osteolepiforms, curves downwards in the anterior part of the lacrimal. This is a consequence of the fact that the fenestra exonarina has migrated ventrally, which has caused modifications of the bones (lateral rostral, maxilla) situated ventrally to that opening. The fenestra exonarina will be considered below, and we may now turn to the other openings of the skull table and include some remarks on the nasolacrimal canal.

The openings of the skull table

Pineal foramen and pineal plates. – As in osteolepiforms, anurans and other eutetrapods with a straight brain stem (see p. 22), the pineal foramen in *Ichthyostega* lies between the frontals. It is a fairly large opening, circular or sometimes (Pl. 11:3) tapering posteriorly into a point. It lies on the top of an elevation shaped like a truncated cone. This pineal elevation is bounded at the base by a more or less distinct groove (Fig. 22D; Pls. 5:2, 3; 11:3), suggesting that the pineal elevation is formed by fused pineal plates. This suggestion is supported by the conditions in osteolepids, in which there is often a median pineal elevation (Fig. 22A; Jarvik 1948a, p. 36), the position and shape of which are similar to those in *Ichthyostega*. Sometimes (Fig. 22A; Jarvik 1948a, p. 58) there are no traces of independent plates, but as is well shown in particular in *Gyroptychius* (Fig. 22B, C; Säve-Söderbergh 1933a, Pl. 14:2; Jarvik 1948a, 1950a, 1985), the pineal foramen is surrounded by independent pineal plates, which, judging from some specimens, may be arranged in an inner and an outer ring. (Fig. 22B; Jarvik 1948a, Pls. 17:3; 30:1, 4). The space occupied by these plates (the pineal fenestra) varies in shape, but in some cases (Fig. 22C; Jarvik 1948a, Pls. 31:6; 32:1; 1950a, Pl. 6:1) it tapers posteriorly into a point and resembles the area bounded by a groove in *Ichthyostega* (Fig. 22D). As evidenced by these facts, the pineal plates in ichthyostegids are retained and have been incorporated in the frontals. As far as I know, no pineal plates have been recorded in post-Devonian tetrapods. However, it is possible that such plates occupied the triangular depression surrounding the pineal foramen in *Peltostega* (Fig. 22E; Wiman 1916, Pl. 15:1; Nilsson 1946), which resembles the tuberculated pineal depression in a similar position in *Osteolepis macrolepidotus* (Säve-Söderbergh 1933a, Pl. 10:3; Jarvik 1948a, p. 58).

Orbital fenestra. – The external opening of the orbit, the orbital fenestra, in *Ichthyostega* (Pls. 1:3; 2; 3; 4:4; 6:1; 8:2; 11:4; 14) is situated a little farther forwards than the pineal foramen. It is usually oval in shape, sometimes (Fig. 10, Pl. 1:3) with 'a curious rectangular incisure' (Säve-Söderbergh

Fig. 22. Pineal foramen and pineal plates. □A–C. The fronto-ethmoidal shield of three osteolepids, (A) *Gyroptychius milleri*, (B) *Gyroptychius agassizi*, and (C) *Gyroptychius milleri* (from Jarvik 1948a, Pls. 29:4; 30:4; 34:3). □D, E, parts of skull table in (D) *Ichthyostega* (A. 194, Pl.5:3) and (E) *Peltostega* (from Wiman 1916, Pl. 15). Terminology after Nilsson 1946, Fig. 1. *om*, margin of orbital fenestra; *s.ot*, sulcus oticus; *s.po*, sulcus postorbitalis; *s.so*, sulcus supraorbitalis; *s.tr.p*, sulcus transversus posterior sensorialis.

1932a, p. 19). In two specimens, A.7 (Fig. 23A; Säve-Söderbergh 1932a, Fig. 10; Pls. 20; 21:2) and A.251 (Fig. 23B, Pl.12:1), it is bipartite, in a way reminiscent of the conditions in the Lower Carboniferous loxommatids (Fig. 23C; Watson 1926; Beaumont 1977). As an explanation to the peculiar key-hole shape of the loxommatid orbital fenestra Romer (1947, p. 94) says that the anterior expansion is generally assumed to be for the reception of some kind of 'glandular structure ', whereas Beaumont (1977, p. 93; Fig. 26) claims that it is caused by 'outbulging of the pterygoides muscle'. According a third theory, recently advanced by Bjerring (1986, p. 32, Fig. 1), 'it housed an electric organ formed from modified eye muscles'.

In loxommatids (Fig. 23C), the anterior border of the orbital fenestra is formed by a bone generally called 'lacrimal', whereas the bone in front of the orbit in other stegocephalians is called 'prefrontal' (Fig. 23D). This condition confounded Säve-Söderbergh. Following his view (1932a) that the bone in front of the orbit in *Ichthyostega* is a compound structure including a posterior 'antorbital' and an anterior supraorbital, he postulated (Säve-Söderbergh 1935a, p. 9) that these two bones in post-Devonian stegocephalians are independent. This had the consequence that the bone in front of the orbit in loxommatids (*La*, Fig. 23C) was called 'posterior "antorbital"' (Säve-Söderbergh 1935a, Fig. 3), whereas the bone in other stegocephalians bordering the orbit anteriorly (*Prt*, Fig. 23D) was called 'anterior supraorbital'. With this terminology arose the question: Where is the lacrimal? Säve-Söderbergh (1935a, pp. 16, 58, 118–120) tried to solve this problem by claiming that the lacrimal had fused with the maxillary into a compound 'lachrymo-maxillary'. This terminology was justifiably criticized by Romer (1936b, p. 535; also Jarvik 1942, p. 548) and is generally abandoned.

Remarks on the nasolacrimal canal. – In one specimen of *Ichthyostega* (Pl. 11:4), the orbital margin of the posterior tectal presents an indistinct opening (*for*), which I suspected could be the opening of the canal for the nasolacrimal duct. However, no traces of the nasolacrimal canal have been observed in any other specimens, and therefore nothing definite can be said about this canal in ichthyostegids. Several years ago (Jarvik 1942, p. 627), when I perused the descriptions of post-Devonian stegocephalians, I found information about the nasolacrimal canal only in a few forms (*Micropholis, Branchiosaurus, Miobatrachus*). It may therefore be of interest to mention that the nasolacrimal canal recently has been described in *Anthracosaurus* (Panchen 1977, p. 460, Fig. 8), in several branchiosaurids (*Apateon, Branchiosaurus*, Boy 1987, p. 78, Figs. 1, 2, 4, 7) and in eryopoids (*Sclerocephalus*, Boy 1988, Figs. 1, 2A; *Acanthostomatops*, Boy 1989, p. 138, Fig. 1; *Onchiodon*, Boy 1990, p. 294, Figs. 1, 2). As is well shown in *Acanthostomatops*, the nasolacrimal canal in larval stages is discernible as a slit on the outside of the lacrimal. This slit, which extends from the 'septomaxillary' to the orbital margin, is soon closed, and in late juvenile stages the canal causes a distinct ridge on the outside of the lacrimal.

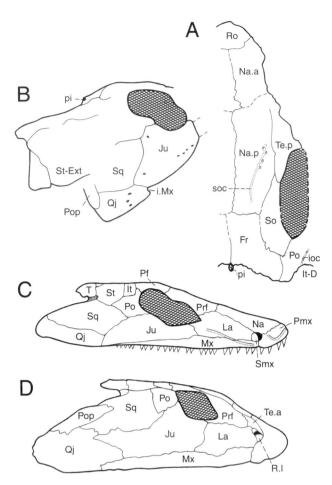

Fig. 23. Orbital fenestra. □A. *Ichthyostega kochi*. Part of skull roof in dorsal view. From Säve-Söderbergh 1932a, Fig. 10. □B. *Ichthyostega*. Part of skull table in posterolateral view (A. 251, Pl. 12). □C. *Baphetes kirkbyi*. Skull table in lateral view. From Beaumont 1977, Fig. 22B. □D. *Crassigyrinus scoticus*. Skull table in lateral view. After Panchen & Smithson 1987, Fig. 11. *Fr*, frontal; *It*, intertemporal; *It-D*, intertemporo-dermosphenotic; *Ju*, jugal; *La*, lacrimal; *Mx*, maxilla; *Na, Na.a, Na.p*, nasals; *Pf*, postfrontal; *Pmx*, premaxilla; *Po*, postorbital; *Pop*, preopercular; *Prf*, prefrontal; *Qj*, qudratojugal; *R.l*, lateral rostral; *Ro*, postrostral; *Smx*, septomaxilla; *So*, supraorbital; *Sq*, squamosal; *St*, supratemporal; *St-Ext*, supratemporo-extrascapular; *T*, tabular; *Te.a, Te.p*, anterior and posterior tectals. *i.Mx*, notch for posterior end of maxilla; *pi*, pineal foramen; *ioc, soc*, infraorbital and supraorbital sensory canals.

However, already in early adult stages this ridge is no longer discernible on the now ornamented outer side of the lacrimal, a condition which probably explains why no distinct traces of the nasolacrimal canal are to be seen in *Ichthyostega* and other ornamented early tetrapods.

Two differences between osteolepiforms and ichthyostegids

In addition to the position of the fenestra exonarina (p. 42) and the presence of peculiar ventral processes of the dermal bones of the skull table (p. 34), the ichthyostegids differ from the osteolepiforms in the following two respects:

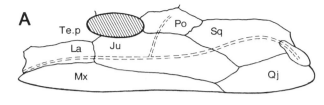

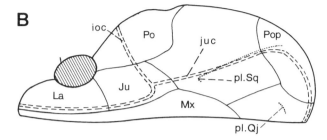

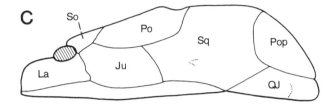

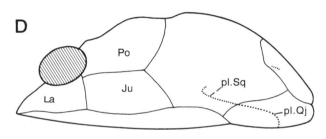

Fig. 24. Diagrammatic representations illustrating the differences in the ventral margin of the orbital fenestra and the posterior extent of the jugal between ichthyostegids and osteolepiforms. □A. *Ichthyostega* (from Fig. 21B). □B. *Eusthenopteron* (from Jarvik 1944, Fig. 18). □C. *Eusthenodon* (from Jarvik 1952, Fig. 26A). □D. *Gyroptychius* (from Jarvik 1948a, Figs. 71, 73). See also Young *et al.* 1992, Fig. 44. *Ju*, jugal; *La*, lacrimal; *Mx*, maxilla; *Po*, postorbital; *Pop*, preopercular; *Qj*, quadratojugal; *So*, supraorbital; *Sq*, squamosal; *Te.p*, posterior tectal; *ioc*, *juc*, infraorbital and jugal sensory canals; *pl.Sq*, *pl.Qj*, squamosal and quadratojugal pit-lines.

1   The ventral boundary of the orbital fenestra (Fig. 24). In the specimens that Säve-Söderbergh referred to *Ichthyostega*, and in all specimens in the new material showing this detail (Pls. 1:3; 2; 3; 4:4; 6:1; 11:4; 12:1), this boundary is formed by the jugal. Anterodorsally, this bone meets the posterior tectal in a short suture, thus excluding the lacrimal from the border of the orbital fenestra. In osteolepiforms, in contrast, the ventral margin of the orbital fenestra is formed partly or wholly by the lacrimal. The orbital margin of the jugal may be short, as in *Eusthenopteron* (Fig. 24B) and in *Canowindra, Beelarongia* and *Koharalepis* (Young *et al.* 1992, Fig. 44). Sometimes the jugal is excluded from the orbit, either by

a contact between the lacrimal and the postorbital, as in *Gyroptychius,* Fig. 24D, or between the lacrimal and the supraorbital, as in *Eusthenodon,* Fig. 24C.

2   In connection with the description of the second difference, which concerns the relations between the jugal and the quadratojugal (Fig. 24), Säve-Söderbergh's quadratojugal canal will be considered.

As illustrated by Säve-Söderbergh (1932a, Fig. 16A, C) and recently emphasized by Panchen & Smithson (1987, p. 360), the jugal and quadratojugal in osteolepiforms are widely separated by the squamosal (Fig. 24B–D), whereas in *Ichthyostega* (and other early tetrapods, Fig. 23C, D) they meet in a suture, thereby excluding the squamosal from contact with the maxilla. What has happened is probably that the jugal has been prolonged backwards by incorporation of squamosal components. The cheek bones are probably compound structures (Jarvik 1980b, pp. 99–100, Fig. 58), but whereas subdivisions of the dermal bones of the skull roof in osteolepiforms are common (Jarvik 1948a; 1980a, Fig. 145), the observed subdivisions of cheek bones are few. A small extra bone (Y) in the posteroventral part of the border of the orbital fenestra has been observed in one specimen of *Eusthenopteron foordi* and one of *Thursius pholidotus* (Jarvik 1944, p. 23, Fig. 10D; 1948a, p. 92, Fig. 67D). Moreover, the postorbital is subdivided in two Australian osteolepiforms (*Canowindra* and *Beelarongia*; Long 1985a, 1987; Young *et al.* 1992). Of interest in this connection is that the squamosal in porolepiforms may be subdivided in various ways (Jarvik 1948a, Figs. 33, 34; 1972, Figs. 41–46). As suggested by Stensiö (1947, p. 142, Fig. 31A–C) and recently discussed and illustrated by Panchen & Smithson (1987, pp. 360–362, Fig. 7) it seems likely that posteroventral squamosal elements – although not yet found independent in osteolepiforms – have been incorporated in the jugal in ichthyostegids and other early tetrapods.

In one specimen of *Holoptychius* (Jarvik 1972, Fig. 46) there is a horizontal row of lower squamosals crossed by the ventral part of the squamosal pit-line, which is continued backwards by the horizontal part of the quadratojugal pit-line. In another *Holoptychius* specimen the squamosal–quadratojugal pit-line is represented by a canal (Jarvik 1948a, Figs. 33; 34). If we imagine that the lower squamosals crossed by a pit-line are incorporated in the jugal, it is readily seen that conditions similar to those found in the Carboniferous tetrapod *Pholiderpeton* (Clack 1987, Fig. 37) are explicable. In that form, as in *Peltostega* (Nilsson 1946, s.qui, Fig. 12), a horizontal sensory-canal groove is found on the posterior part of the jugal and the anterior part of the quadratojugal. This is of interest, because Säve-Söderbergh (1932a, p. 71, Figs. 11, 12) claimed that he, on the single specimen of *Ichthyostegopsis wimani* 'after a fortunate preparation', had found 'traces of sensory canals on the quadratojugal, probably belonging to a quadratojugal canal (*qc*, Fig. 11) homologue of the vertical pit-line of the cheek in fishes' (the term 'quadratojugal canal' was introduced in a paper then under preparation; see Säve-Söderbergh 1933a, p. 8). The photographs of *Ichthyostegopsis* were taken before the 'fortunate preparation', and traces of the quadratojugal canal are therefore not visible on his plates (1932a, Pls. 16, 1; 17:1; 18:1, 2), nor are they shown on the specimen as preserved today. However, as noted above, the filling of the sensory canals disclosed by Säve-Söderbergh is fragile. It is thus possible that the traces of a horizontal quadratojugal canal which, as pointed out by Säve-Söderbergh (1932a, p. 71), 'by no means are so clear and undoubted as the remains of the jugal and infraorbital canals' have been lost through the years. Be this as it may, no horizontal quadratojugal canal as figured by Säve-Söderbergh (1932a, Figs. 11, 12, 15) has been found in the new material.

However, in one specimen of *Ichthyostega* (A. 251, Fig. 23B; Pl. 12:1) which, as many other specimens, presents distinct sensory-canal pores on the anterior part of the jugal, there is in addition a vertical row of three pores on the anterior part of the quadratojugal and a less distinct pore on the posterior part of the jugal. This tallies well with the fact (Jarvik 1967, Fig. 10B, D) that the cheek of *Palaeoherpeton* ('*Paleogyrinus*') as restored by Panchen (1964, Fig. 12), shows a sensory-canal groove running much as the continuous squamosal–quadratojugal pit-line in *Osteolepis macrolepidotus* (Säve-Söderbergh 1933a, 1941; Jarvik 1948a, Figs. 21A, 22) or *Gyroptychius agassizi* (Fig. 24D) and like that crossing the anterior part of the quadratojugal. However, judging from the position of the pores, the canal in *Ichthyostega* is a jugalo-quadratojugal canal.

Neural endocranium and associated exoskeletal processes and laminae

As established by mechanical preparations and saw cuts, the neural endocranium in *Ichthyostega* is incompletely preserved. This is true in particular of the ethmoidal region and the anterior part of the orbitotemporal region. Moreover, the otic region is strongly influenced by strange exoskeletal laminae and processes of the skull table which border or penetrate this region. Because of these conditions, and since unfortunately no skull with lower jaws suitable for investigations with Sollas's grinding method has been found, a detailed description of the neural endocranium cannot be given. Also it has turned out to be difficult to interpret with confidence the canals that pierce its walls.

As in osteolepiforms and other 'crossopterygians', the neural endocranium in ichthyostegids includes ethmosphenoid and otoccipital divisions. These two divisions are continuous dorsally, underneath the dermal cranial roof, whereas ventrally they are separated by a narrow transverse slit, the *fissura preoticalis* (Figs. 25, 36, 37), which is a vestige of the osteolepiform intracranial joint (Jarvik 1980b, p. 67, Figs. 41, 42). The fissura was first figured, but not named, by Säve-Söderbergh (1932a, Pl. 17:2) in *Ichthyostegopsis* (A. 5,

Pl. 22:3). In the new material the fissura preoticalis is well shown in palatal view in several other specimens (Pls. 4, 19–21, 26, 27, 28). Moreover, it is shown in longitudinal sections in two specimens (Pls. 16:2–4; 21:1, 2).

Owing to the fragility of the skeleton, the skull table of some specimens, e.g., A. 158 (Pls. 14, 15), may be removed, exposing the underlying neural endocranium. The brittleness of the skeleton and the configuration of the palatoquadrate often cause the endocranium to break into two portions close behind the basipterygoid process but in front of the fissura preoticalis.

The anterior portion, including the anterior part of the orbitotemporal region and the ethmoidal region (Pl. 15), in addition to the opening of the endoskeletal pineal canal shows only traces of the dorsal part of the ethmosphenoid without any structural details. This is true also of the bluish periosteal lining of the lateral wall of the endocranium shown in some longitudinally split specimens. Because of this poor preservation the anterior part of the orbitotemporal region and the ethmoidal region in *Ichthyostega* remain practically unknown. (As regards the basipterygoid process and the basal articulation, see p. 40.)

The posterior portion of the broken endocranium, which extends forwards to include the posterior part of the orbitotemporal region with the fossa hypophysialis, as seen in dorsal view, is dominated by the hemispherical otic elevations (Pls. 15; 19:2; 23:1; 25:1). Between these elevations a dorsal median part of the endocranium is preserved, but it is generally too damaged by compression from the sides to show any details. Lateral to the median strip of endoskeleton, the dorsal side of the otic elevation shows a bluish periosteal lining, indicating that this part has been covered by a thin layer of endoskeletal bone. It may be noticed that fossa bridgei (or fossa supra-auditiva, Bjerring 1984a, p. 232), in contrast to conditions in osteolepiforms (Säve-Söderbergh 1936, Figs. 58, 59; Romer 1937, Figs. 1, 2; Jarvik 1954, 1980a; Vorobyeva 1977, and others) and many post-Devonian stegocephalians (Säve-Söderbergh 1936; post-temporal fossa, Romer 1947; Beaumont 1977; Smithson, 1985; and others), is lacking in *Ichthyostega*, as it is in embolomeres and probably in *Crassigyrinus* according to Panchen (1985, p. 514). The broken dorsal ends of exoskeletal processes are shown on the dorsal side of the otic elevation. The fragile bony tissue of these processes may be easily removed, which results in deep holes or spaces leading down to the palatal side.

The most important of these processes is the 'curious ventral process of the lateral extrascapular' (supratemporo-extrascapular) described by Säve-Söderbergh (1932a, p. 24, Fig. 3) and most conveniently referred to as the *Säve-Söderbergh process* (*pr.S.-S*; Figs. 35, 37). The dorsal opening of the space for that process is seen at the posterolateral corner of the otic elevation in several specimens and, as is well shown in A. 152 (Pl. 23:1, 3, 4), it may be crescent-shaped. The ventral part of the Säve-Söderbergh process is visible in palatal view in several specimens (Pls. 16:2; 20:1, 2; 23:2;

Fig. 25. Ichthyostega. Restoration of palate. For explanations, see Figs. 28C and 37.

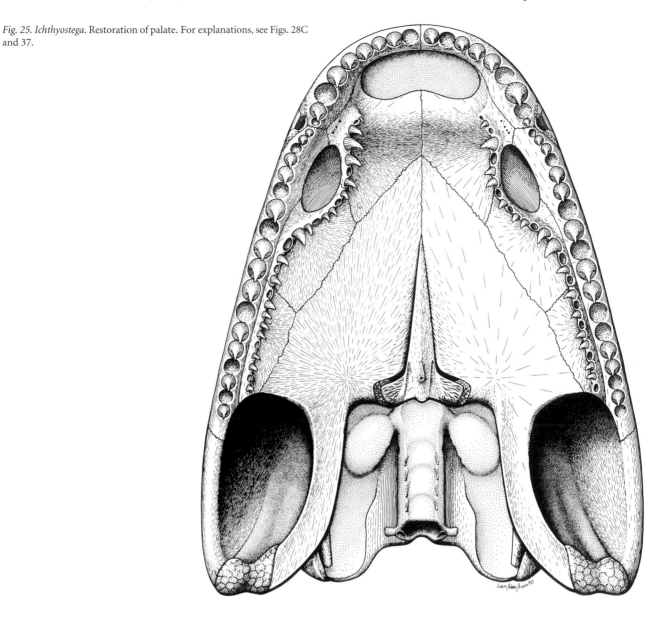

26:2), as well as in *Ichthyostegopsis* (Pl. 22:1, 3; Säve-Söder-bergh 1932a, Pl. 17:2), but is best shown in A. 55 (Pl. 26:1) and A. 117 (Pl. 25:2, 3). As evidenced by the latter two specimens, the ventral part of the process includes an antero-medial rod-like portion (*pr.S.-S.r*) and a posterolateral lami-nar portion (*pr.S.-S.la*). The rod-like portion widens in its ventral part to form a supporting area, which abuts against the dorsal side of a flange (*fl.Enpt*) of the quadrate ramus of the entopterygoid (p. 44). Posteriorly the process is em-braced by a strong lamina (*la.Sq*) of the squamosal. This lamina is bent forwards and conceals the main part of the laminar portion of the process. Ventral processes or crests of the supratemporo-extrascapular (tabular) are found also in post-Devonian stegocephalians as, e.g., in *Bronthosuchus* and *Thoosuchus* (Bystrow & Efremov 1940, p. 126, Fig. 13; Getmanov 1989, Fig. 16), but as far as I know, the strange

conditions encountered in *Ichthyostega* are unparalleled in other tetrapods.

In A. 152 (Pl. 23:1) the broken dorsal end of another exoskeletal structure is seen on the posteromedial part of the otic elevation. The space (*sp. long. cr.*) occupied by that structure leads down to the longitudinal palatal crest (*long.cr*, Pl. 23:2) described below (p. 36). It could not be ascertained if the anterior end of the exoskeletal part of this crest occu-pied the space and extended upwards to the skull table.

The deep pit (*sp.pr.Fr*) shown in A. 152 (Pl. 23:1) close in front of the otic elevation was occupied by crushed exoskel-etal bone, probably belonging to a ventral process of the skull table. This is supported by the conditions in two other specimens. In the corresponding place in A. 117 (Pl. 25:1) lies the crescent-shaped opening of a space (*sp.pr.Fr*) leading down (Pl. 25:2) to the area anterolaterally to the bluish vesicle in the palate described below as the sacculus vesicle. More-over, in A. 158 (Pl. 16:2) a detached process (*pr.Fr*) is found in the same area. This process is suggestive of the rod-like ventral part of the Säve-Söderbergh process seen farther back in the same specimen, a condition which suggests that the space in A. 117 was occupied by a ventral process, most likely of the frontal. It could not be established if the ventral widened end also of that process abutted against the flange of the quadrate ramus of the entopterygoid.

A noteworthy fact is also the presence of a broad and deep depression or groove in the posterolateral part of the otic elevation. This groove, the *parotic groove* (*gr.pot*, Pls. 19:1, 2; 20:1, 2; 23:3, 4; 24:1) is occupied by a strange medial *parotic ridge* of the palatoquadrate (*ri.pot*, Fig. 27; Pls. 15, 23:5; 24:2). Dorsally the parotic groove extends upwards to, or almost to, the dorsal opening of the space for the Säve-Söderbergh process, but in some specimens (see below) that opening is found in the bottom of the dorsal part of the parotic groove.

Close behind the dorsal end of the groove for the parotic palaquadrate ridge the otic elevation presents a 'rounded knob' mentioned by Säve-Söderbergh (1932a, p. 66) and ventral to that a broad ridge. This ridge (Pls. 19:2; 22:1; 23:1) has a bluish periosteal lining, and some specimens show that it was covered by a thin endoskeletal lamina. On the lateral side of that lamina, a narrow canal (*can*, Pls. 19:2; 22:1; 23:1, 3–5) runs downwards from the area of the 'knob' to the ventral margin of the endocranium. The thin endoskeletal lamina is strengthened on the outside by the 'oblique poste-rior lamina of the lateral extrascapular' (supratemporo-extrascapular) described by Säve-Söderbergh (1932a, p. 66). The broad ridge is discernible on the posterior (exoskeletal) side of the vertical *complex lamina* thus formed (*vert.la*, Fig. 35, Pl. 12:1, 2). Dorsally this lamina is bounded by a distinct edge at the posterior margin of the ornamented dorsal side of the supratemporo-extrascapular. Medial to that edge the posterior part of the parieto-extrascapular presents a median depression. Laterally the depressed area on each side is pierced by the canal for the occipital artery (*c.a.occ.*). Similar conditions are found also in A. 64 (Pl. 3), and the part of the depressed area situated between the canals *c.a.occ* is obvi-ously the area of attachment for the supraneural ligament (*ar.sn.li*, Fig. 35). In A.251 also the area for the attachment of the supradorsal ligament (*ar.sd.li*, Fig. 35) is discernible immediately dorsal to the posterior opening of the cranial cavity representing the foramen magnum (see p. 51). The large posteriorly facing area ventral to the edge of the supra-temporo-extrascapular in A.251 (Pl. 12:1, 2) obviously served as attachment area for the trunk musculature. Medi-ally the complex vertical lamina curves gently forwards, and its concave anterior side forms the posterior boundary of a characteristic *troughlike longitudinal depression* seen in ven-tral (palatal) view (Fig. 37). In that depression the complex lamina (vert.la) continues forward as a *longitudinal crest* (*long.cr*) extending to the bluish sacculus vesicle described below. Since the space (sp.long.cr) seen in the posteromedial

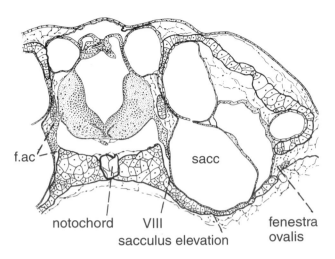

Fig. 26. *Xenopus laevis*. Transverse section through the otic region of a 16 mm larva. From Paterson 1949, Fig. 8. *f.ac*, foramen acousticum; *sacc*, sacculus; VIII, n.acousticus.

part of the otic elevation in A. 152 (Pl. 23:1), as mentioned above (p. 35), leads down to this area, it seems likely that the anterior end of the exoskeletal part of the longitudinal crest occupied the space and was joined to the skull roof. Posteri-orly the medial side of the longitudinal crest is joined to the dorsolateral process (*pr.dl*, Fig. 35) of the otoccipital. The notch (*i.X*, A. 117, Pl. 25:2, 3) on the ventral side of the connection thus formed is presumably for the passage of the vagus nerve.

The bluish vesicle referred to above is more or less well preserved, and it is shown in many specimens (Pls. 16:1, 2, 5; 19:1; 20:1, 2; 21:4; 22:1, 3; 23:2; 25:2, 3). It is formed by a thin bluish film, which represents the periosteal lining of a part of the otoccipital. On the outside it was covered by a layer of fragile endoskeletal bone, remains of which in some speci-mens are preserved at the margin of the vesicle. In the present state of preservation, the vesicle is empty (Pl. 16:5), but in Recent frogs (*Xenopus*, Fig. 26 herein, Paterson 1949; *Rana*, Gaupp 1893, 1904, Figs. 153–160) it housed the sacculus of the membranous labyrinth and has therefore been termed the *sacculus vesicle*.

Anteromedially the sacculus vesicle is continuous with the median tubular portion of the otoccipital, whereas postero-medially in all specimens it is separated from that portion by a gap widening posteriorly. In this gap two foramina are seen in the lateral wall of the braincase. The foremost and larger of these two foramina (VIII, Pls. 19:1, 3, 4; 25:2) leads into the sacculus vesicle (Pl. 16:5), and most likely it represents the foramen acusticum housing the acoustic ganglia (cf. Fig. 26; Gaupp 1904, pp. 704–705, Figs. 160, 161). Because the poste-rior of the two foramina and other foramina in the wall of the braincase (Pls. 19:3, 4; 20:3, 4, 5) cannot be identified safely, I have not attempted to interpret these structures. Since no other opening of the sacculus vesicle is discernible, it may be tentatively suggested that the fenestra ovalis was situated in the gap just mentioned. This suggestion is supported by the fact that the vestibular fontanelle with the fenestra ovalis in osteolepiforms (*Eusthenopteron*, Jarvik 1980a, Figs. 88B, 93; 1985, Figs. 26B, 27A; *Spodichthys*, Jarvik 1985, Figs. 26A, 27B) is situated in a similar position to the gap in *Ichthyostega*. Moreover, a rod-like structure (*rod*) possibly representing a part of the stapes is seen in the gap in some specimens of *Ichthyostega* (Pls. 19:1; 20:1). Unfortunately, the stapes, which presumably reaches the otic notch (p. 31) is too incompletely preserved to provide any reliable information regarding details of this bone, and also the canals for the semicircular ducts and other details in the structure of the membranous labyrinth remain unknown in ichthyostegids.

A prominent structure in the palate (Fig. 37) is the tubular median portion of the otoccipital shown in palatal or posterior views in several specimens (Pls. 4:1–3; 16:1, 2, 5; 19:1; 20:1, 2; 21:3, 4; 22:1, 3; 23:2; 24:3, 4; 25:2, 3; 26:2; 27:2; see also Säve-Söderbergh 1932a, Pl. 17:2) and in longitudinal sections in two specimens (A. 62: Pl. 21:1, 2; A. 158, Pl. 16:2, 4).

In a previous restoration (Jarvik 1952, Fig. 36) it is said that the large posterior opening of the otoccipital is the opening for the notochordal canal. This is not quite true, since this large opening in its dorsal part includes the posterior opening of the cranial cavity (foramen magnum). However, since the tubular portion is composed of cranial vertebrae it will be treated in connection with the description of the vertebral column (p. 51).

Palatoquadrate and basal articulation

The palatoquadrate in *Ichthyostega* is completely ossified, as it is in some specimens of *Eusthenopteron* and other osteolepiforms known in this regard (Jarvik 1954; 1972, Fig. 26). In consequence of the flattening of the skull, the palatoquadrate in *Ichthyostega* (Fig. 27), as well as in later stegocephalians, is twisted, and we may distinguish between an anterior horizontal or palatal part and a posterior vertical, or pterygoid, part including the pars quadrata (Säve-Söderbergh 1936, pp. 117–118; Romer 1947, p. 30).

The horizontal or palatal part is best shown in A. 158 (Pls. 17, 18) in which it, however, is represented mainly by the bluish dorsal periosteal lining (*per.li*). The low space between this lining and the dorsal side of the underlying dermal bones of the palate shown in the counterpart (Pl. 17:1) were occupied by a thin endoskeletal lamina. Most of the fragile endoskeletal bone tissue of this lamina was lost when the skull was split, but remains are retained most posteriorly (*end.pq*) in the thickened area of the basal process. Judging from the extent of the bluish lining, the horizontal palatal lamina anteriorly reached the ethmoidal region, and presumably it was joined to that region in some way.

A conspicuous feature of the dorsal side of the horizontal palatal part of the ichthyostegid palatoquadrate (A.158, Pl. 18) is the presence of branching grooves that housed branches of nerves and blood-vessels (Fig. 27B). We may

distinguish a main posteromedial groove, which divides into two grooves, one (*gr.r.md*) running in posterolateral and one (*gr.r.mx*) in anterolateral direction. The proximal part of the main posteromedial groove turns slightly upwards, and obviously it lodged a structure that passed through the notch (*i.V*) between the ascending and paratemporal processes. A similar notch is present also in *Eusthenopteron* and was in this form traversed by the common maxillo-mandibular trunk of the trigeminal nerve, which on the dorsal side of the palate divided into a ramus maxillaris running forwards and a r.mandibularis running backwards (Jarvik 1980a, Fig. 129). Judging from these conditions, it is obvious that the main posteromedial groove in *Ichthyostega* housed the common maxillo-mandibular trunk, which divided into a r.maxillaris running forwards and a. r.mandibularis running backwards in the grooves described above. However, the notch in *Eusthenopteron* was also traversed by the arteria temporalis, which divided into anterior and posterior branches (a.maxillaris, a.mandibularis) on the dorsal side of the palatoquadrate (Bjerring 1968, Fig. 4; Jarvik 1980a, Fig. 131), and most likely these arteries were present also in *Ichthyostega*.

Posteriorly, at the transition from the horizontal (or palatal) to the vertical (or pterygoid) part, the palatoquadrate in *Ichthyostega* bends inwards to form a characteristic transverse vertical wall (*tr.w*). The anterior side of that wall (Fig. 27 F), well shown in A. 5 (Pl. 22:2; also A.115, Pl. 21:5, A. 139, Pl. 24:1), presents the *processus ascendens palatoquadrati* (*pr.asc*). This process projects from the ventromedial part of the transverse wall upwards and forwards to the dermal cranial roof, which it reaches in the posterolateral part of the frontal, close to the suture between that bone and the intertemporo-dermosphenotic. Accordingly, the processus ascendens in *Ichthyostega* has about the same relations to the dermal cranial roof as in *Eusthenopteron* and the Triassic stegocephalian *Lyrocephalus* (Jarvik 1980b, Fig. 127). Anteroventrally the ascending process is separated from the horizontal palatal part by a notch (*i.v.io*), presumably for the passage of the vena infraorbitalis which is a tributary to the jugular vein (cf. *Eusthenopteron*, Bjerring 1968, Fig. 4; *Amia*, Jarvik 1980a, Figs. 19, 64).

Behind the ascending process, as in *Eusthenopteron* (Jarvik 1954, Fig. 16; 1980a, Fig. 107) and post-Devonian stegocephalians (Säve-Söderbergh 1936, Fig. 9; Romer 1947), follows the notch (*i.V*) for the maxillo-mandibular trunk of the n. trigeminus. This notch is bounded posteriorly by the *processus paratemporalis* (*pr.pt.*), which in post-Devonian stegocephalians generally is called *processus oticus* but was recently termed *paratemporal process* in the embolomere *Pholiderpeton* (Clack 1987, p. 21, Fig. 4). In *Ichthyostega*, the paratemporal process reaches the dermal cranial roof somewhere in the area of the centre of radiation of the intertemporo-dermosphenotic. A remarkable fact is that the process dorsally, as in *Eusthenopteron* (Jarvik 1954, p. 27, Fig. 12), is expanded into a lateral, almost horizontal plate (*hor.pl*). In *Eusthenopteron* (Jarvik 1980a, p. 150, Figs. 107A,

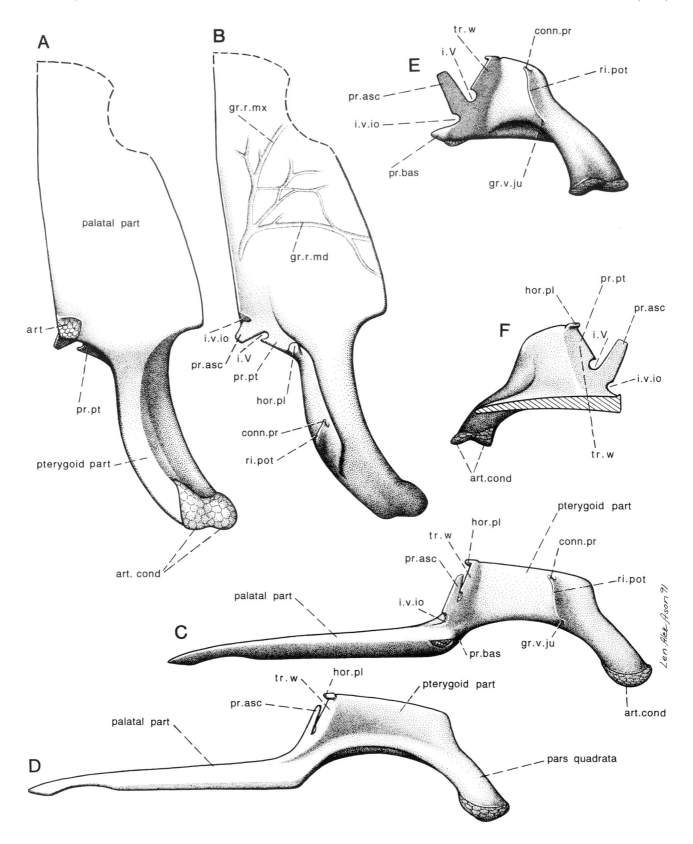

Fig. 27. Restorations of the palatoquadrate of *Ichthyostega* in (A) ventral, (B) dorsal, (C) medial, (D) lateral, (E) posterior, and (F) anterior views. *art*, articular area of basal process; *art.cond*, articular condyles of pars quadrata; *conn.pr*, connecting process; *gr.r.md*, *gr.r.mx*, grooves for r.mandibularis and r.maxillaris V; *gr.v.ju*, groove of for jugular vein; *hor.pl*, horizontal plate of paratemporal process; *i.v.io*, groove for r.infraorbitalis of jugular vein; *i.V*, notch for n.trigeminus; *pr.asc*, processus ascendens; *pr.bas*, processus basalis; *pr.pt*, paratemporal process; *ri.pot*, parotical ridge or crest; *tr.w*, transverse wall.

108, 118), this plate was probably joined by a ligament to the ventral side of the anterior part of the intertemporal.

At the lateral end of the transverse wall, the vertical ptery-goquadrate part of the palatoquadrate bends distinctly backwards and continues, close to the otic elevation, to slightly behind the deep parotic groove in the lateral side of that elevation (see below). Dorsally this part of the palatoquadrate, from the distinct bend backwards, extends to the dermal cranial roof in the part formed by the intertemporo-dermosphenotic and most posteriorly by the squamosal.

A remarkable characteristic of the ichthyostegid palatoquadrate is that it is provided with a strong medial ridge, the *parotic ridge* (*ri. pot*), which occupies the deep *parotic groove* in the lateral side of the otic elevation. Ventrally the *parotic ridge* presents a groove, *gr.v.ju*, running forwards (Pl. 24:2). As will be explained below in connection with the discussion of the origin of the parotic ridge, this groove was developed for the passage of the jugular vein. Dorsally the parotic ridge extends to the dorsal opening of the space for the ventral process of the supratemporo-extrascapular (the Säve-Söderbergh process) where it is, as shown in A.152 (Pl.23:3), joined to the dorsal part of the Säve-Söderbergh process by a connecting process *(conn.pr)*. That there is connection between the parotic ridge and this ventral process is supported by A. 63, in which the space for the ventral process opens into the bottom of the dorsal part of the groove for the ridge.

Since thus the parotic ridge occupies the deep parotic groove in the lateral side of the otic elevation and, moreover, is joined to the Säve-Söderbergh process, it is obvious that the palatoquadrate in *Ichthyostega* is firmly anchored to the braincase. The skull in ichthyostegids, as in osteolepiforms, is thus akinetic, which is also suggested by the presence of a rigid skull table, attached to the braincase by strong descending processes and laminae, and by the structure of the palate. The widespread view about intracranial kinetism in osteolepiforms is a delusion unsupported by reliable data. As a matter of fact, *Eusthenopteron*, studied with the aid of the Sollas Grinding Method on the basis of excellent fossil material, is one of the few osteolepiforms in which the problem of the intracranial kinetism can be penetrated (Jarvik 1937, pp. 112–117, Fig. 18; 1944, pp. 30–37, Figs. 14, 15; 1948a, p. 13;1954, pp. 27–32; 1980a, pp. 108–111, 146–150, 163–173; Bjerring 1967, p. 234; 1971, p. 191; 1972, p. 89; 1978, p. 205; 1985, p. 36).

On the origin of the parotic ridge. – In order to interpret this prominent and, as far as I know, unique structure among early tetrapods, we may first turn to *Eusthenopteron* and Recent anurans (Jarvik 1972, pp. 215–217). A characteristic feature of *Eusthenopteron* and other teleostome fishes is the presence of a lateral commissure situated outside the jugular vein (see, e.g., Jarvik 1980a, Figs. 14, 17–19, 86A, B, 88B, 249B). It has been argued (Jarvik 1954, pp. 73–75) that this structure represents the suprapharyngohyal (cf. Bjerring 1993, p. 134) which has been incorporated into the neural

endocranium. It may be noticed that the lateral commissure, as is well shown in *Eusthenopteron* (Jarvik 1980a, Fig 76C, 86A), has established a broad dorsal connection with the anterior part of the crista parotica.

In tetrapods no distinct lateral commissure is discernible but, as maintained by van der Westhuizen (1961, pp. 62–63), in Recent anurans it is probably represented by the so-called processus oticus of the palatoquadrate. However, judging from observations by van Eeden (1951) in larval stages of the anuran *Ascaphus* and the conditions in *Eusthenopteron*, it seems likely that the processus oticus, as tentatively suggested in a previous paper (Jarvik 1972, p. 217, Fig. 90) 'is a double formation consisting of a medial part which is the lateral commissure (processus dorsalis in anurans, van Eeden 1951, p. 104, Figs. 29, 32, 34B, 35) and a lateral part formed by the palatoquadrate. This condition will arise if we assume that the palatoquadrate in *Eusthenopteron* – after the reduction of the anterior part of the spiracular gill-slit – fuses with the adjoining side of the lateral commissure. That this is what really has happened is supported also by the fact that the processus dorsalis in anurans may be separated from the palatoquadrate by a layer of connective tissue (van Eeden 1951, p. 110, Fig. 35) and is further evidenced by the fact that the squamosal has retained its position in relation to the palatoquadrate.' This is supported by the Swanepoel's statement (1970, p. 101) that the crista parotica in another anuran (*Breviceps*) 'has a dual origin, being derived from the otic process and from the top end of the hyoid arch'. Since 'the top end of the hyoid arch' obviously is the dorsal end of the lateral commissure (dorsal process, van Eeden) it follows that the lateral commissure in anurans is joined dorsally to the crista parotica, as it is in *Eusthenopteron*.

In view of these facts it is to be concluded that the parotic ridge in *Ichthyostega* is the lateral commissure, which has fused with the palatoquadrate in the way I suggested in 1972. This conclusion is supported by the presence of a distinct groove (*gr.v. ju*) on the inner side of the ventral part of the parotic ridge (Fig. 27C, E; Pl. 24:2), which probably carried the jugular vein. Moreover, the part of the palatoquadrate in *Ichthyostega* that carries the parotic ridge dorsally reaches the squamosal part of the dermal cranial roof.

Quadrate portion and jaw joint. – A little behind the parotic ridge, the palatoquadrate in *Ichthyostega* decreases markedly in height and continues backwards to the jaw joint (A. 139, Pl. 24:1–3). The anterior part of the low quadrate portion (*pars quadrata*) is covered ventrally and medially by the quadrate ramus of the entopterygoid, whereas the more solid posterior part with the transversely placed articular surface generally is well exposed (Pls. 23:2, 24:1–3, 26:1–2, 27:1–2). The articular surface (*art.cond*, Figs. 27, 37) situated between the quadratojugal and the entopterygoid, displays medial and lateral knobs separated by a shallow groove and matches the articular surface of the lower jaw. Accordingly, the jaw joint in *Ichthyostega* agrees with that joint in osteolepiforms

(*Eusthenopteron,* Jarvik 1944, Fig. 12; 1980a, Figs. 107, 124, 125), porolepiforms (Jarvik 1972, Figs. 30, 47) and post-Devonian stegocephalians (Romer 1947, p. 42).

In most specimens of *Ichthyostega* (Pls. 4: 1; 24:3; 25:2; 26:2), the jaw joint lies slightly behind the level of the posterior end of the median tubular part of the otoccipital. The fact that in A. 55 (Pl. 26:1) it is found in front of that level is probably due to compression.

Basal process and basal articulation. – The basal process of the palatoquadrate in *Ichthyostega* is in most specimens found together with the basipterygoid process of the ethmosphenoid (Pls. 18; 20:1; 23: 1; 25:11; 26:1, 2; 28; 29). Both processes forming the basal articulation are covered ventrally and posteriorly by dermal bones, the basal process by the entopterygoid and the basipterygoid process by the ascending process of the parasphenoid (Fig. 37). The two processes are therefore generally concealed. However, the articular surfaces are well displayed on both processes in A. 234 (Pl.28). The basal process (see also A. 149, Pl. 29) shows two articular areas meeting at right angles, an anterior one facing posteromedially and a shorter posterior one facing anteromedially. Matching articular surfaces also meeting at right angles are found on the basipterygoid process. A characteristic feature is that the two processes generally are separated by a wall of the embedding matrix (see Pls. 26:2; 28). In the living animal, the space between the two processes was probably occupied by cartilage covering the articular surfaces, and conceivably the cartilage sheaths were joined by synchondrosis, permitting only a slight flexibility at the basal articulation.

Dorsal to the basal articulation is a wide canal (*c.v.ju.* A. 64, Pl. 4:2; A. 158, Pl.18), no doubt for the jugular vein. After receiving the infraorbital tributary (see p. 37), this vein ran backwards dorsal to the basipterygoid process and continued in the groove on the inside of the parotic ridge.

Dermal bones and openings of the palate

General remarks. – A characteristic feature of teleostome fishes and tetrapods, in contrast to dipnoans and other plagiostomes, is the presence of two dental arcades, outer and inner, both in the upper and lower jaws (Jarvik 1980b, pp. 92–94). In the upper jaw, the outer arcade is formed by two bones (premaxilla and maxilla) and the inner by three bones (vomer, dermopalatine and ectopterygoid).

All the tooth-bearing bones that form these arcades are carried by elements of the visceral arches and belong to the visceral exoskeleton. On the basis of new facts, provided by the grinding series of *Eusthenopteron* and data from the anatomical and embryological literature, it was possible to identify the three dorsal elements (infrapharyngeal, suprapharyngeal and epal) characteristic of the independent visceral arches also in the hyoid, mandibular and premandibular arches (Jarvik 1954). Later, it has been demonstrated (Bjerring 1977) that there is an additional visceral arch, the terminal one in front of the premandibular. It is important

to note that independent dental plates in the ground skull of *Eusthenopteron* were retained in their natural position, not only on the independent visceral arches (including the palatoquadrate) but also on the neural endocranium. I have hypothesized that the neural endocranium is composed of cranial vertebrae, to which certain dorsal elements of visceral arches have been added (Jarvik 1959b, 1960, 1972, 1980a, 1980b). On the basis of these concepts, it has now become possible to present definitions of the dermal bones of the palate in terms of their position relative to the visceral arch elements and the axial portion of the neural endocranium (Jarvik 1972, 1980b, pp. 90–100, 247–248, Figs. 48–49, 53–56).

One of the two bones in the upper outer dental arcade, the maxilla, is formed by modified lateral epipremandibular dental plates (Jarvik 1980b, Fig. 53), whereas the premaxilla most likely is developed in relation to the epal element of the terminal arch and represents modified epiterminal dental plates. The three bones in the inner upper dental arcade are formed by horizontal (ventral) dental plates of the infrapharyngeal (vomer) and epal (dermopalatine and ectopterygoid) elements of the premandibular arch. The large bone following inside the inner dental arcade, the entopterygoid, is developed in relation to the commissural lamina (Jarvik 1972, p. 70, Fig. 25; 1980b, pp. 79, 92, Figs. 46, 47, 49). In this connection it may be emphasized that the entopterygoid and the outer dental arcade are lacking in dipnoans and other plagiostomes, as is also a choana. For these and other obvious reasons, the unfortunate and misleading term 'Sarcopterygii' is to be rejected (Jarvik 1980a, pp. 400–403, 427, 436; 1980b, pp. 257–258; regarding the terms *Osteolepipoda* and *Urodelomorpha,* see Jarvik 1980b, p. 149, Fig 140; 1986, p. 14; 1988; pp. 39–40). In dipnoans (and other plagiostomes) only the inner upper dental arcade is present and constitutes the 'pterygoid' and the palatine tooth plate, which include equivalents to the ectopterygoid, dermopalatine and vomer in teleostomes. As in the latter, the tooth-bearing bones of the inner dental arcade in dipnoans are carried by the epal and infrapharyngeal elements of the premandibular arch, which, however, in dipnoans, in sharp contrast to the conditions in teleostomes, have been incorporated in the neural endocranium and form the 'pseudotrabecle' and one half of the trabecle (palatoquadrate) commissure (Bjerring 1977, Fig, 31; Jarvik 1980a, pp. 401–403).

The parasphenoid, present also dipnoans and certain other plagiostomes, arises in relation to the part of the neural endocranium that is formed by cranial vertebrae (Jarvik 1980b, pp. 95–98).

Comparisons between osteolepiforms (*Eusthenopteron*), *Ichthyostega and later stegocephalians*

A remarkable condition at the transition from fish to tetrapod is the shift in the main biting function from the inner to

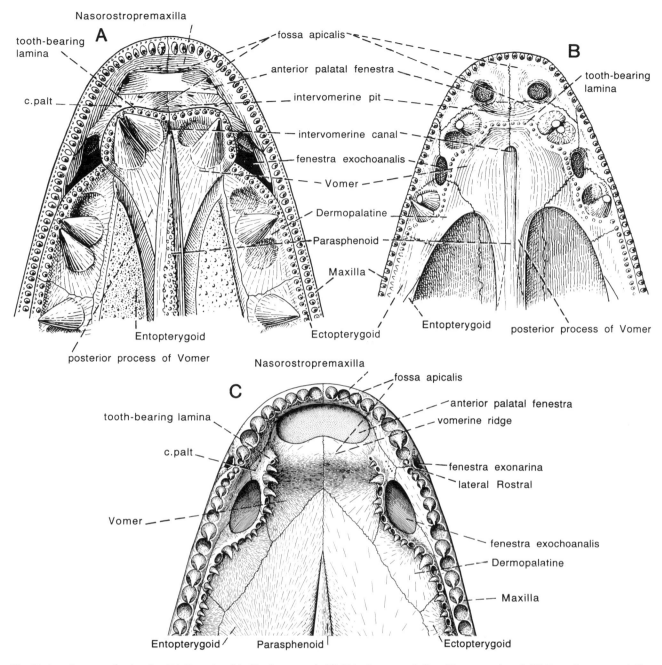

Fig. 28. Anterior part of palate in (A) Devonian fish *(Eusthenopteron)*, (B) Triassic stegocephalian *(Heptasaurus)*, and (C) Devonian stegocephalian *(Ichthyostega)*. A, B from Jarvik 1942 (Fig. 87); C, anterior part of Fig. 25. *c.palt*, canals for terminal twigs of r.palatinus VII.

the outer dental arcade (Jarvik 1980b, Fig. 54). In osteolepiforms (and porolepiforms) the dermal bones in the inner dental arcade carry large, sharp tusks. In *Ichthyostega* the teeth in the outer arcade are relatively stronger than in *Eusthenopteron*, and although some of the teeth in the inner arcade may be rather large, real tusks are lacking (Pls. 26:1; 27:1). It is of interest then, that such tusks are present in post-Devonian stegocephalians, and most remarkable are the conditions in Triassic stegocephalians such as *Heptasaurus*. As I have repeatedly emphasized (Jarvik 1942, pp. 632–637,

Fig. 87; 1955a, Fig. 8; 1964, p. 66, Fig. 22; 1980b, p. 223, Fig. 137), *Heptasaurus* (Fig. 28B) is suprisingly similar to *Eusthenopteron* (Fig. 28A) in the structure of the anterior part of the palate; more similar even than to *Ichthyostega* (Fig. 28C) and other Palaeozoic stegocephalians. As in *Eusthenopteron*, there is thus in *Heptasaurus* a well-defined fossa apicalis bounded posteriorly by a transverse tooth-bearing lamina of the vomer. Also, an intervomerine pit and an intervomerine canal are present, as well as vomerine tusks. Moreover, the vomerine tooth-bearing lamina has a rostro-caudal portion, which borders the fenestra exochoanalis. Also as in *Heptasaurus*, the vomer is provided with a posterior process situated along the parasphenoid, a bone which extends far forwards underneath the ethmoidal region. In *Ichthyostega*, in contrast, the vomer lacks a posterior process, and the entopterygoids meet in a median suture, a condition that has caused a reduction of the anterior part of the parasphenoid. Moreover, the transverse tooth-bearing lamina of the vomer, which in *Eusthenopteron* and *Heptasaurus* bounds the fossa apicalis posteriorly, is replaced in *Ichthyostega* by a low transverse vomerine ridge (Pls. 11:1. 26:1. 27:1, 29). This ridge is usually toothless, but teeth may be present in is lateral part. In front of the transverse ridge the vomer is continued by a thin laminar part. Such a thin part is present also in *Eusthenopteron*, and in both the fish and the tetrapod it forms the posterior boundary of the anterior palatal fenestra, which is bordered anteriorly and anterolaterally in a similar way by the palatal laminae of the naso-rostro-premaxillae.

In *Eusthenopteron*, the ventral side of the ethmoidal region of the neural endocranium is exposed in the anterior palatal fenestra, and laterally the openings of two important canals are found (Fig. 28A). In *Ichthyostega*, remains of the endoskeleton may be seen in the fenestra (Pl. 11:1), but no openings of canals are discernible. However, close to the suture with the naso-rostro-premaxilla, the vomer is pierced by a row of foramina (*c.palt.* Fig. 28C; Pls. 4:1; 10:1; 11:1; 17:2; 26:1; 32:3). Judging from the conditions in *Eusthenopteron* (Fig. 28A; Jarvik 1942, Figs. 56, 60; 1980a, Fig. 128C, D) these foramina probably transmitted terminal twigs of the r. palatinus VII.

The rostro-caudal portion of the vertical tooth-bearing lamina of the vomer is retained in *Ichthyostega* (Fig. 28C; Pls. 11:1; 26:1; 27:1; 29) and, as in *Eusthenopteron* (Fig. 28A), it forms the anterior part of the medial margin of the fenestra exochoanalis. Also as in *Eusthenopteron*, this fenestra is bordered posteromedially by the anterior part of the vertical tooth-bearing lamina of the dermopalatine, which is continued posteriorly by the vertical tooth-bearing lamina of the ectopterygoid. In *Eusthenopteron*, the anterior tip of the dermopalatine overlaps the vomer (Jarvik 1942, p. 456, Figs. 54A, 56, Pl. 12:3; cf Jarvik 1981, p. 38) as it does at least in certain stegocephalians akin to *Heptasaurus* (*Benthosuchus*, Bystrow & Efremov 1940, p. 123, Fig. 2; *Sassenisaurus*, Nilsson 1942, Fig 1; Jarvik 1942, p. 635). However, in contrast to conditions in *Eusthenopteron*, the horizontal lamina of the dermopalatine of *Ichthyostega*, *Heptasaurus* and several other stegocephalians (Säve-Söderbergh 1935a; Case 1946; Romer 1947; Watson 1962) is joined to the vomer in a suture (Fig. 28B, C). The dermopalatine in *Ichthyostega* is also provided with a lateral lamina sutured to the maxilla, forming a groove leading backwards from the fenestra exochoanalis.

An important difference between *Eusthenopteron*, on one hand, and *Ichthyostega* and other eutetrapods, on the other, concerns the anterior extent of the maxilla. As pointed out previously (Jarvik 1942, pp. 497, 548), the maxilla in most extant anurans reaches much farther forwards than in *Eusthenopteron*, and in consequence the premaxilla is relatively short, not reaching the fenestra exochoanalis. In *Eusthenopteron*, there is a broad ridge between the fossa apicalis and the fenestra exochoanalis formed by the naso-rostro-premaxilla and the vomer, and the fenestra exochoanalis is bordered laterally by the naso-rostro-premaxilla and maxilla. In *Ichthyostega*, there is a similar ridge, but the ridge is formed partly by the maxilla which extends farther forwards than in *Eusthenopteron*. It forms a suture with the vomer, excluding the naso-rostro-premaxilla from the fenestra exochoanalis, which is bordered laterally by the maxilla. The same is true also of *Heptasaurus* (Fig. 28B) and some other stegocephalians.

Fenestra exochoanalis. – This opening in *Ichthyostega*, described by Säve-Söderbergh (1932a) in four specimens (A.1; A.3; A.4; A.5), is shown also in Nr 220 (Pl. 1:1, 2), collected by Kulling in 1929 and in several specimens in the new material (Pls. 4:1, 2; 7:1; 11:1; 17:2; 26:1; 27:1; 29; 33:3; see also Pl. 22:3). In most respects it agrees well with that in *Eusthenopteron* but, as may be gathered from the statements above, it is bordered by three bones (vomer, dermopalatine, maxilla) whereas in *Eusthenopteron* the naso-rostro-premaxilla shares in its boundary also (Fig. 28).

Fenestra exonarina and lateral rostral. – Before proceeding to *Ichthyostega* it may be convenient to recall the conditions in osteolepiforms, particularly *Eusthenopteron*, in which the snout is well-known (Jarvik 1942; 1966, Fig 13C; 1980a). In *Ichthyostega*, the fenestra exonarina has a different position from that in osteolepiforms (Jarvik 1966, Fig. 14), but the view maintained by Westoll (1937, p. 28; 1943, p. 89, Fig. 8A) that it may be situated at the very margin of the frontoethmoidal shield and confluent with the fenestra exochoanalis is a mistake based on compressed and insufficiently prepared fossil material (Jarvik 1942, pp. 485–486, Figs. 62B, 63; 1948a, pp. 36, 60). In *Eusthenopteron*, the fenestra exonarina (Jarvik 1942, Figs. 34, 52B, 53, 68A; 1966, Figs. 8, 13C, 14A) is bordered dorsally by the anterior tectal, which is provided with an inwardly directed process, and ventrally by the lateral rostral with its processus dermintermedius. The infraorbital sensory canal passes straight forward through the lacrimal, the lateral rostral and the posterior part of the naso-rostro-premaxilla to be continued by the ethmoidal commissure (Jarvik 1944, Figs. 17, 18; 1980a, Figs. 116A, B, 127A, B). Below the lateral rostral lies the naso-rostro-premaxilla, posterior to which lies the maxilla. These three dermal bones, together with the overlying part of the neural endocranium (the lamina nariochoanalis, Jarvik 1942, p. 245; 1980a, p. 111, or epiterminal, Jarvik 1980b, p. 86), form a strong bridge, the *pons nariochoanalis* (Jarvik 1942, p. 246; 1986, p. 3), separating the fenestra exonarina from the fenestra exochoanalis. This bridge is an important passage for the infraorbital sensory canal, nerves and blood vessels (see, e.g.,

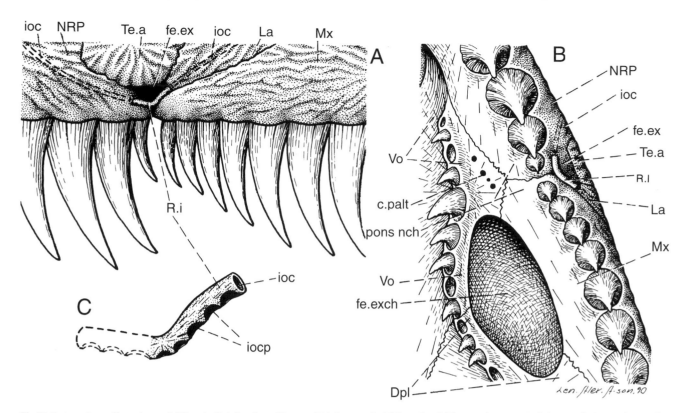

Fig. 29. Restorations of lateral rostral (C) and adjoining dermal bones of *Ichthyostega* in (A) lateral and (B) ventral aspects. *Dpl*, dermopalatine; *La*, lacrimal; *Mx*, maxilla; *NRP*, nasorostropremaxilla; *R.l*, lateral rostral; *Te.a*, anterior tectal; *Vo*, vomer. *c.palt*, canals for terminal twigs of r.palatinus VII; *fe.ex*, fenestra exonarina; *fe.exch*, fenestra exochoanalis; *ioc*, infraorbital sensory canal; *iocp*, pores of infraorbital canal; *pons nch*, pons nariochoanalis.

Jarvik 1942, Figs. 53, 58, 60). Moreover, it is to be noticed that in *Eusthenopteron* the upper outer dental arcade is straight outside the fenestra exochoanalis, as it is in other choanate fishes and post-Devonian tetrapods.

Turning to the ichthyostegids, we find (as noticed above, p. 31) a wide gap in the lateral side of the skull table (Pls. 4:4; 9:2; 10:2; 11:1; 17:2, 3; see also Säve-Söderbergh 1932a, Pls. 11; 12:1; 13). This gap, which has been termed the *exonarial gap*, is bordered dorsally by the anterior tectal, anteriorly by the naso-rostro-premaxilla and posteriorly by the antero-ventral corner of the lacrimal and the ornamented part of the maxilla. The unornamented inturned anterior part of the maxilla is visible in the gap.

Of great interest is the tubular bone found in the posterior part of the exonarial gap on the left side in A. 64 (Pl. 10). This bone (Fig. 29) was briefly described in a previous paper (Jarvik 1952, footnote 1, pp. 80–81) and was called 'lateral rostral', although it only includes the sensory-canal portion of the osteolepiform lateral rostral. The fate of the processus dermintermedius in *Ichthyostega* is unknown, and no traces of an inwardly directed process of the anterior tectal have been found. The preserved part of the tubular lateral rostral measures about 16 mm in length. The dorsal part of the bone bends slightly backwards, and its dorsal end is found close to the anteroventral corner of the lacrimal. In contrast to conditions in *Eusthenopteron*, the infraorbital sensory canal in

Ichthyostega curves ventrally in the anterior part of the lacrimal and is at the anteroventral corner of the bone directly continued by the canal in the tubular lateral rostral. That this really is a sensory-canal bone is evidenced by the fact, not mentioned in 1952, that the ventrolateral side of the tube is pierced by a row of rather large foramina (*iocp.*, Fig. 29) of about the same size as the pores of the infraorbital canal in the lacrimal in a specimen (A.4), of *Ichthyostega* dissected by Säve-Söderbergh (1932a: Pls. 11; 12:2). Close to the antero-ventral end the lateral rostal bends distinctly forwards ,and the tube is directed towards the posterior opening of the infraorbital sensory canal in the naso-rostro-premaxilla. Obviously the bone has been broken a little in front of the bend and its anterior part has been tentatively restored. The reason for the break at just this place is easily accounted for if we consider the position of the bone and the conditions in the specimen (A.64) in which it has been found. The bend is situated dorsal to the suture between the naso-rostro-pre-maxilla and the maxilla. As seen in palatal view, these bones are generally joined in sutures to each other and to the vomer. However, in A. 64 (Pl. 10:1) they lie apart, and the breakage of the lateral rostral in this place is obviously due to postmortal dislocations.

The fenestra exonarina in *Ichthyostega* (*fe.ex.* Fig. 29) is, as in *Eusthenopteron*, situated dorsal to the lateral rostral. Dorsal to the pons nariochoanalis (*pons nch*), which is formed by the naso-rostro-premaxilla, the maxilla and the lateral rostral, it is in wide communication with the fenestra exochoanalis (cf. A. 55, Pl. 26:1; note that the pons on the left side was torn off when the cast was taken away from the specimen). As demonstrated long ago (Jarvik 1942, pp. 622, 649), the ventral position of the external nostril in *Ichthyostega* cannot be a primitive condition as maintained by Säve-Söderbergh (1932a, pp. 98–99). As is quite evident from the facts presented in previous paragraphs, it is due to a secondary ventral migration, which has led to modifications of the lateral rostral and other structures at the margin of the mouth. As far as known, the lateral rostral is represented only by its sensory-canal component. As a consequence of the ventral shift, the infraorbital sensory canal in the anterior part of the lacrimal curves downwards, and passing the lateral rostral it makes a sharp bend upwards to continue without interruption in the canal in the naso-rostro-premaxilla (cf. Fig. 30). In this regard, we find similar conditions as in other forms in which the anterior or posterior nostrils have been displaced (porolepiforms and urodeles, Jarvik 1972, pp. 153, 195, 196, 271, Figs. 7, 36C, 43–45, 79D, E, 107B, C; dipnoans and sharks, Jarvik 1980a, Fig. 310). Another consequence of the ventral migration is that the anterior part of the maxilla has been driven inwards. The upper outer dental arcade formed by the maxilla and the naso-rostro-premaxilla therefore describes a curve inside the external nostril in a way unparalleled in later tetrapods. This unique specialization further debars the ichthyostegids from close relationship to later tetrapods and emphasizes their isolated position.

Entopterygoid. – As described by Säve-Söderbergh (1932a, p. 82) the entopterygoid in *Ichthyostega* (Figs. 25, 28C, 35, 37) includes an anterior palatal and a posterior quadrate ramus. Behind the anterior median suture, backwards to the basal process, the medial marginal part of the palatal ramus is bent upwards. This part, which on its medial side presents a series of delicate oblique ridges (*stri. Enpt*, Pl. 29), forms the lateral boundary of the narrow interpterygoid vacuity (*ipter. vac,* Pls. 21:4; 27:1). The two foramina in my previous restoration (Jarvik 1952, Fig. 36) are shown only in one specimen (A. 64, Pl. 4:1, 2) and are omitted in the new restoration (Fig. 25). At the basal process (*pr.bas*, Pls. 20:1; 21:4; 26:1, 2; 28, 29), the entopterygoid forms an angular notch with one margin along the posteromedially facing part of the articular area of the basal process and one along the anteromedially facing area. Behind the slightly raised rectangular corner of the notch (A. 53, Pl. 26:2) the entopterygoid bends upwards, and together with a part of the parasphenoid it forms a vertical transverse wall, which bounds the trough-shaped longitudinal palatal depression anteriorly. Laterally this ascending lamina bends backwards to the quadrate ramus, where forms the lateral boundary of that depression (Fig. 37). In contrast to conditions in *Eusthenopteron,* the ventral part of the quadrate ramus, situated between the trough-like depression and the adductor fossa, is flattened. This flattened part projects medially into a distinct flange (*fl. Enpt*, Fig. 35, Pls.

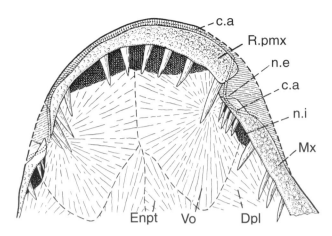

Fig. 30. Ichthyostega eigili. Restoration of a part of the skull, seen obliquely from the anterior, ventral and left side to show position of external naris in relation to the dermal bones. *Dpl*, dermopalatine; *Enpt*, entpterygoid; *Mx*, maxillary; *R.pmx*, rostropremaxillary; *ca*, probable course of anterior commissure of the infraorbital canals; *n.e*, external naris; *n.i*, internal naris. Drawing and explanations from Säve-Söderbergh 1932a, Fig. 8.

25:2, 3; 26:1). As mentioned above (p. 35) the rodlike portion of Säve-Söderbergh's ventral process abuts against the dorsal side of the flange.

Parasphenoid. – The parasphenoid in *Ichthyostega* (Figs. 25, 28, 35, 37; Pls. 4:1, 2; 17:1; 20:1; 21:3, 4; 22:3; 23:1; 24:3, 4; 26:1, 2; 27:1, 2; 28:1) in several respects resembles that in *Eusthenopteron* but is relatively shorter than in that form (Fig. 28A). Anteriorly it reaches only to the posterior part of the median suture between the entopterygoids and does not extend forwards underneath the ethmoidal region. Posteriorly it extends to the fissura preoticalis, which is a vestige of the osteolepiform intracranial joint. Accordingly it extends exactly as far backwards as in *Eusthenopteron* and as in that form it expands posteriorly into a bilateral process. This process is homologous to the processus ascendens anterior in *Eusthenopteron* and other fish, and although in *Ichthyostega* – because of the flattening of the skull here as in other early tetrapods – it is almost horizontal in position, it most conveniently may be referred to by the same name as in *Eusthenopteron.*

As is well established (Jarvik 1954, p. 59), the *processus ascendens anterior* in fishes is formed by modified infrapharyngeal dental plates of the mandibular visceral arch. A remarkable fact is that the closely set denticles of the palatal dermal bones in *Eusthenopteron* gradually may be replaced by a system of low anastomosing ridges (Jarvik 1937, p. 89; 1944, p. 38). In *Ichthyostega* (see e.g., A. 234, Pl. 28), such ridges are present on the outside of the processus ascendens. Posteriorly the process bends sharply upwards and, covering the posterior side of the basipterygoid process, it forms the medial part of the anterior transverse wall of the troughlike palatal depression (Fig. 37). Laterally this ascending lamina of the parasphenoid meets the ascending lamina of the entopterygoid. At its dorsal margin is a notch, which forms the ventral boundary of a wide canal leading forward. This canal certainly transmitted the jugular vein (*c.v.ju*) which, as in *Eusthenopteron*, passed backwards dorsal to the basipterygoid process.

As in *Eusthenopteron*, the parasphenoid in *Ichthyostega* is provided with a tooth-bearing ridge (*ri. P*), pierced by the buccohypophysial canal (*f.bh*. Fig. 37; Pls. 20:1; 21:2–4; 22:3; 26:2; 27). However, this ridge is relatively shorter than in *Eusthenopteron* and, moreover, the denticles have been modified and are discernible as granules or delicate ridges.

Lateral to the tooth-bearing ridge, the ventral side of the parasphenoid presents a distinct depression widening backwards. This depression is directed towards the shallow pit on the anterior part of the tubular portion of the otoccipital (Pls. 21:4; 26:2; 27:1, 2;28). The development and position of the depression and the pit (*bcr. mu*, Fig. 37) strongly suggest that a small paired basicranial muscle, somewhat as restored by Bjerring (1971, Fig. 6) in the palaeoniscid *Pteronisculus*, was retained in ichthyostegids (see also Tatarinov 1994).

Also in post-Devonian stegocephalians, the parasphenoid is provided with a processus ascendens anterior ensheathing the basipterygoid process (Romer 1947, p. 43). The fact that the parasphenoid in many fishes and tetrapods extends more or less far backwards underneath the otical and occipital regions is due to incorporation of dental plates corresponding to those found independently in the grinding series of *Eusthenopteron* and *Glyptolepis* and represented by more or less independent primordia in larval stages of fishes and extant amphibians (Jarvik 1954, pp. 37–70; 1972, pp. 83–84; 1980b, pp. 95–98). Also the prolonged posterior part of the parasphenoid may be provided by an ascending process. As revealed by the grinding series of *Eusthenopteron* and other evidence, this process, the *processus ascendens posterior*, is formed by fused and modified dental plates representing ascending infrapharyngeal dental plates of the hyoid arch (Jarvik 1954, p. 57). Since this process is present in post-Devonian stegocephalians (Romer & Witter 1942, p. 940; Romer 1947, pp. 42–47; Beaumont 1977, p. 63) as well as in extant anurans (Gaupp 1896; Stadtmüller 1936), it is likely that such dental plates are present also in ichthyostegids, and it may be predicted that they will be found if a well-preserved and complete skull is investigated with the Sollas Grinding Method.

The lower jaw

In the structure of the lower jaw, *Ichthyostega* (Fig. 31) agrees strikingly with *Eusthenopteron* (Jarvik 1937, 1944, 1980a, Figs. 125, 127, 129. 131, 132, 1980b, Fig. 97; 109). The ornamented external side is more or less well shown in several specimens (Pls. 1:2; 6:1; 12:1; 13:3–5; 30:1; 32:1; 33). As in *Eusthenopteron*, *Eusthenodon* (Jarvik 1952, Fig. 27B), *Megalichthys* (Watson 1926, Fig. 37), *Osteolepis* (Jarvik 1948a, Fig. 40 M), and *Latvius* (Gross 1956, Fig. 32; Jessen

1966, Fig. 2B, Pl. 7:1), this side is composed of five bones, a dorsal dentary and a ventral series of four bones, which I since 1944 have termed *infradentaries*. Because of the ornamentation, the sutures may be indistinct, but as evidenced by several specimens they run very much as in *Eusthenopteron*. Also in post-Devonian stegocephalians (see, e.g., Romer 1947, pp. 60–61, Figs. 9, 10; Jupp & Warren 1986) there is a dorsal dentary and a ventral series of four bones given various names (for synonyms see Nilsson 1944, pp. 17–19). The foremost of these bones (infradentary 1) is thus sometimes called 'splenial', sometimes 'presplenial'. The next bone (infradentary 2) is generally called 'postsplenial'. However, when the first bone is called 'presplenial' and the second 'postsplenial' (Nilsson 1943, 1944; Panchen 1985, Fig. 12) it may be justified to ask where the splenial has gone. The bone following behind the postsplenial (infradentary 3) is generally called 'angular', whereas the hindmost bone (infradentary 4) by some writers (Säve-Söderbergh 1935a, Bystrow 1938, Fig. 19; Bystrow & Efremov 1940; Gross 1941, Figs. 26, 27; Getmanov 1989; Shishkin 1973) is called 'supraangular', and still others (e.g., Watson 1926, Fig. 10; 1962; Romer 1947, p. 61, Figs. 9, 10; Beaumont 1977, Fig 14; Carroll 1988, p. 141) use the name 'surangular'.

A remarkable agreement between *Ichthyostega* and *Eusthenopteron* is that, in contrast to conditions in post-Devonian stegocephalians, the dentary reaches backwards to the jaw joint. Consequently, the lateral dorsal margin of the adductor fossa in *Eusthenopteron* and *Ichthyostega* is formed by the dentary, whereas this margin in post-Devonian stegocephalians is formed by the bone called 'supra-angular' or 'surangular'. In this respect, the ichthyostegids have retained the primitive condition, whereas in the dentition they have reached the tetrapod stage. In *Ichthyostega*, (Fig. 31; Pls. 31:2;.32:2; 31:1, 2), as in *Eusthenodon* (Jarvik 1952, Pl. 11:2; 1972, Fig. 49) and several other osteolepiforms (Jarvik 1966, p. 57), the dentary is provided with an enlarged tooth anteriorly (*de. tusk*). A noteworthy fact is, however, that in *Ichthyostega* the teeth of the dentary as well as those in the upper outer dental arcade are relatively stronger than the corresponding teeth in *Eusthenopteron* (Figs. 28A, C). This condition, together with the fact that tusks in *Ichthyostega* are lacking on the bones of the inner dental arcade both in the lower and upper jaws, demonstrates that the shift in the main biting function from the inner to the outer dental arcades has taken place in *Ichthyostega*. In this respect, the ichthyostegids are more advanced than post-Devonian stegocephalians in which the vomer and dermopalatine (Fig. 28B), and often also the ectopterygoid, bear tusks (Romer & Witter 1942, Fig. 3; Beaumont 1977, p. 38, Figs. 9, 23; Smithson 1982, Fig. 11; Boy 1989, Fig. 2; Getmanov 1989, Figs. 6, 12).

The sensory lines of the lower jaw in *Ichthyostega*, either enclosed in canals opening outwards with superficial pores or displayed by narrow grooves, are more or less well shown in several specimens (Pls. 13:3, 5; 30:1, 3; 32:1, 3; 33). The course of the sensory lines in *Ichthyostega* agrees with that in

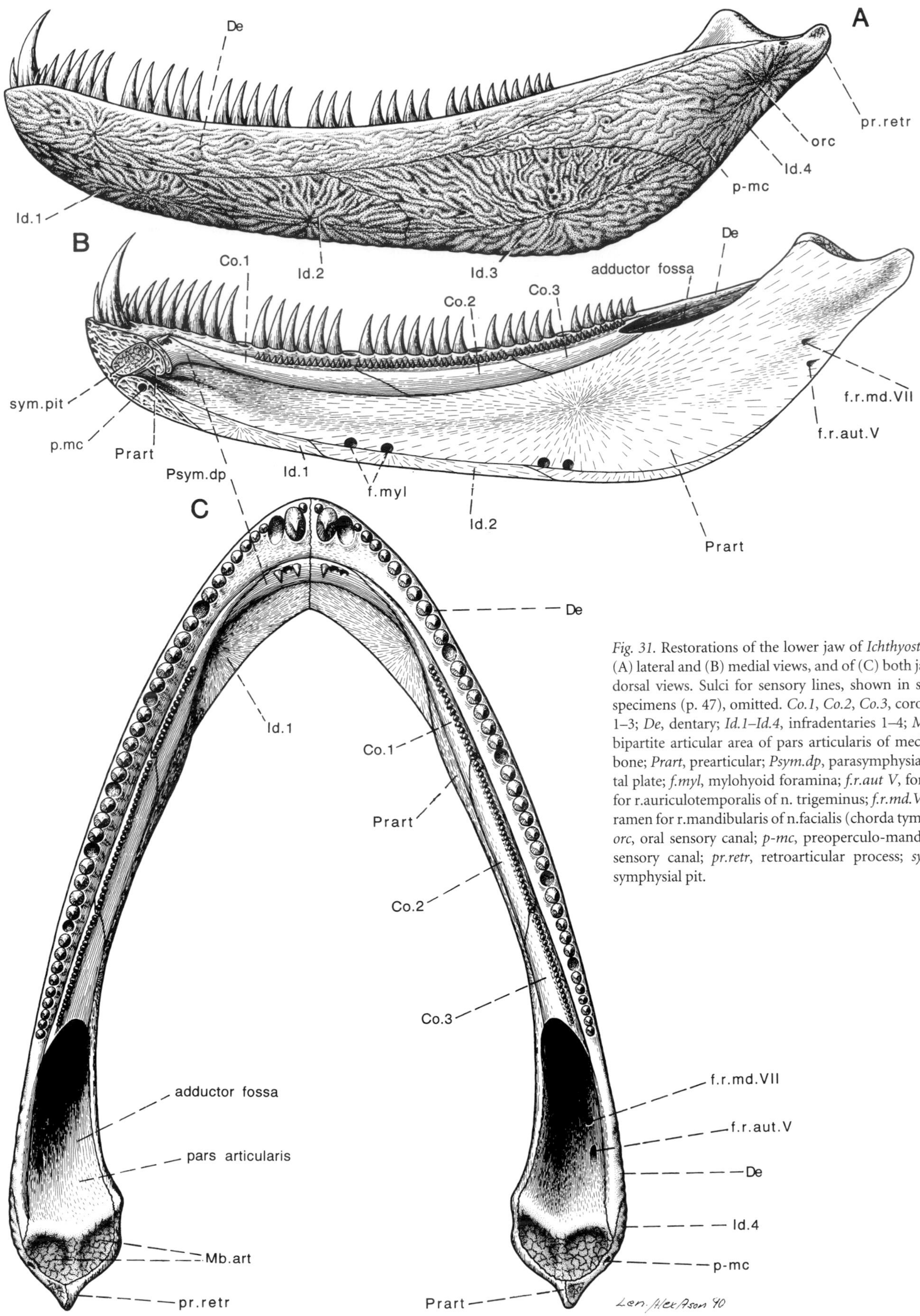

Fig. 31. Restorations of the lower jaw of *Ichthyostega* in (A) lateral and (B) medial views, and of (C) both jaws in dorsal views. Sulci for sensory lines, shown in several specimens (p. 47), omitted. *Co.1, Co.2, Co.3,* coronoids 1–3; *De,* dentary; *Id.1–Id.4,* infradentaries 1–4; *Mb.art,* bipartite articular area of pars articularis of meckelian bone; *Prart,* prearticular; *Psym.dp,* parasymphysial dental plate; *f.myl,* mylohyoid foramina; *f.r.aut V,* foramen for r.auriculotemporalis of n. trigeminus; *f.r.md.VII,* foramen for r.mandibularis of n.facialis (chorda tympani); *orc,* oral sensory canal; *p-mc,* preoperculo-mandibular sensory canal; *pr.retr,* retroarticular process; *sym.pit,* symphysial pit.

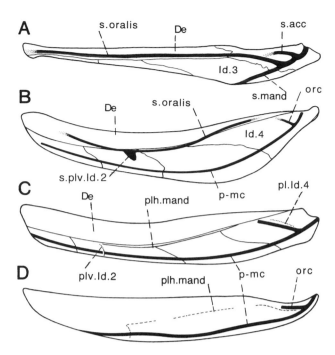

Fig. 32. Diagrammatic representations of sensory lines, pit-lines or open grooves (sulci) in the lower jaw. ☐A. *Benthosuchus*, from Bystrow 1935 (Fig. 14). ☐B. *Ichthyostega* (see Pls. 32:1, 33). ☐C. *Eusthenopteron*, from Jarvik 1944 (Fig. 11). ☐D. *Osteolepis*, from Jarvik 1948a (Fig. 32). *De*, dentary; *Id.3, Id.4*, infradentary 3 and 4; *orc*, oral sensory canal; *p-mc*, preoperculo-mandibular sensory canal; *pl.Id.4*, pit-line of infradentary 4; *plh.mand*, horizontal mandibular pit-line; *plv.Id.2*, vertical pit-line of infradentary 2; *s.acc*, sulcus accessorius; *s.mand*, sulcus mandibularis; *s.plv, Id.2*, depression for vertical pit-line of infradentary 2; *s.oralis*, sulcus oralis.

Eusthenopteron: in both, the mandibular portion of the pre-operculo-mandibular canal pierces the four infradentaries in a similar way, and in both there is a short oral canal in infradentary 4 (*Eusthenopteron*, Jarvik 1944, Fig. 11; *Thursius*, Jarvik 1948a, Pl. 23:1, 2). However, in *Eusthenopteron* (and other osteolepiforms) there are also mandibular pit-lines. Before discussing their fate in *Ichthyostega* it is convenient to turn to the post-Devonian stegocephalians (Fig. 32), in which the sensory lines are lodged in a system of branching open grooves (Bystrow 1935, p. 96; 1938, p. 248; Bystrow & Efremov 1940; Nilsson 1943; 1944, pp. 7–11). Bystrow distinguishes one ventral branch, which, as is now well known (*Megalocephalus*, Beaumont 1977, p. 67, Fig. 14; *Colosteus*, Hook 1983, p. 20), runs on the outside of the elements of the infradentary series from the area of the jaw joint to the symphysis, where it is continued by its fellow. This branch, called 'sulcus marginalis' by Bystrow and 'sulcus mandibularis (sensorialis)' by Nilsson, evidently lodged a sensory line corresponding to the mandibular portion of the preoperculo-mandibular canal in *Ichthyostega* and *Eusthenopteron*. Moreover, there is a short groove, the sulcus accessorius (*s.acc*, Fig. 32A), which obviously corresponds to the oral sensory canal in ichthyostegids and some osteolepiforms. However, in addition to these branches there is a conspicu-ous dorsal branch (*s. oralis*) termed 'sulcus dentalis' by Bystrow and 'sulcus oralis' by Nilsson. This groove (Bystrow 1935, Fig. 35; 1938, Fig. 6; Bystrow & Efremov 1940, Fig. 29; Nilsson 1943, Figs. 2, 10, 13, 20; 1944, Figs. 2, 4, 8, 29, 33; Shishkin 1973, Figs. 40. 44, 61, Pl. 4: 5) passes forwards on the outside of the supra-angular (infradentary 4) and continues more or less far forwards on the dentary, close to the suture with the angular (infradentary 3).

According to Nilsson (1944, p. 10) 'the lateral line grooves of the Stegocephalian lower jaw are certainly comparable to the canals and pit-lines of the sensory line system in Crossop-terygians'. He was obviously of the opinion that the sulcus oralis in stegocephalians is represented by the mandibular pit-line in 'crossopterygians', and in a footnote he says that 'in the Stegocephalians no equivalent has been observed so far of the vertical part of mandibular pit-line of Crossopterygians'.

The mandibular pit-line consisting of a long horizontal and a short transverse or vertical part is well shown on the cosmine-covered external side of the lower jaw in many osteolepids (Fig. 32D, Säve-Söderbergh 1933a, 1941; Jarvik 1948a, 1949, 1950a, 1985; Gross 1956; Jessen 1966; Vorobyeva 1977; Young *et al.* 1992). Of special interest are the conditions in *Eusthenopteron* (Jarvik 1937, p. 89, Fig. 9; 1944, p. 45) in which the position of the mandibular pit-line in relation to the dermal bones could be made out. It has been established (Fig. 32C) that the vertical part (*plv.Id.2*) is situated on infradentary 2, whereas the horizontal part runs in the suture between the dentary and infradentaries 2 and 3 and that most posteriorly it crosses infradentary 4 running along the oral sensory canal.

These facts demonstrate that the posterior part of the sulcus oralis (dentalis) in post-Devonian stegocephalians running on the supra-angular (infradentary 4) must be equivalent to the pit-line of infradentary 4 in *Eusthenopteron* and the corresponding part of the mandibular pit-line in the osteolepids. The remaining part of the sulcus oralis, which runs on the dentary, posteriorly close to the suture to the angular, is evidently equivalent to the horizontal part of the mandibular pit-line in osteolepiforms. The fact that it runs on the dentary, and not as in *Eusthenopteron* in the suture between the dentary and the infradentaries, is certainly due to increase in height of the dentary at the cost of the underlying infradentaries, which has been shown to be an evolutionary trend in stegocephalians (Nilsson 1944, pp. 57, 62).

After the discussion of the sulcus oralis in post-Devonian stegocephalians it may be of interest to record that the outer side of the lower jaw in one specimen of *Ichthyostega* (Pl.33) presents a slightly depressed, unornamented area. This area or groove (s. oralis), more or less well-shown also in other specimens (Pls. 32:1; 33:2–4), follows the suture between the dentary and the infradentaries from the jaw joint to the posterior part of infradentary 1, running partly on the dentary, partly on the infradentaries. As is readily seen, this unornamented area or groove runs very much as does the sulcus oralis in post-Devonian stegocephalians and must be its homologue. Accordingly it may be concluded that the mandibular pit-line was retained in *Ichthyostega*, presumably embedded in a thickened part of the skin that occupied the groove and caused the obliteration of the ornamentation. Judging from some specimens, also the vertical pit-line of infradentary 2 (*s.plv.Id.*2, Pls. 32:1; 33:3) was retained.

The coronoid series in *Ichthyostega* (Fig. 31; Pls. 30:2, 3; 31: 1–3; 32:2), as in some osteolepiforms, includes four elements. Although less specialized than in porolepiforms (Jarvik 1972, pp. 113–120, Figs. 47–50) and struniiforms (Jessen 1966, Fig. 10), the foremost element has been termed *parasymphysial dental plate* in osteolepiforms, and this name may be used also in *Ichthyostega*. In osteolepiforms this plate is still unknown in *Eusthenopteron foordi* (tentatively restored by Jarvik 1972, Fig. 50B) but has been described in many other forms (Vorobyeva 1962, 1975, Fig. 3, 1977, 1980; Vorobyeva & Lebedev 1986; Jessen 1966, p. 331, Fig. 5; Young *et al.* 1992, Fig. 18). In *Ichthyostega*, the parasymphysial dental plate and the following three coronoids are composed of a medial laminar part and a marginal low ridge carrying small teeth. Coronoid tusks are lacking. A remarkable fact is that teeth are lacking on the anterior part of coronoid 1, which is wedged between the dentary and the parasymphysial dental plate. The latter plate (*Psym.dp*) carries three teeth, of which generally one or two are preserved. Moreover, a characteristic feature of the parasymphysial dental plate in *Ichthyostega* is that its laminar part is bent strongly downwards, covering the anterior end of the prearticular (Fig. 31B; Pl. 32:2).

Until quite recently, only three coronoids (*Co 1– Co 3*), given various names (Nilsson 1944), have been known in post-Devonian stegocephalians. It is therefore of interest that Holmes (1989, p. 181, Fig. 15) has found a parasymphysial plate in the Lower Permian antracosauroid *Archeria* and also reported it from a Carboniferous stegocephalian. The fact that this plate in *Archeria* is 'wedged between the dentary and splenial' is certainly due to an evolutionary trend involving a shortening of the prearticular ('gonial') in post-Devonian stegocephalians (Nilsson 1944, pp. 60–61).

The prearticular (*Prart*) in *Ichthyostega* (Fig. 31B, C; Pls. 24:1–3; 30:2, 3;31; 32:3) extends from the retroarticular process (*pr.retr*), of which it forms the medial half, to the symphysial pit (*sym.pit*) in which the anterior end of the meckelian bone is exposed. Thus it extends exactly as far forwards as in *Eusthenopteron* and bounds the symphysial pit in the same way (Jarvik 1937, 1972, Fig. 49). No traces of dentition have been observed, a condition that is readily accounted for since the dentition of the prearticular in *Eusthenopteron* is carried by an independent dental plate (Jarvik 1937, pp. 89, 112).

In addition to some very small foramina (Pl. 31:2, 3), the inner side of the prearticular presents two groups of large foramina (*f. my*) situated close to the ventral margin and partly on the adjoining infradentaries. Corresponding foramina are found also in osteolepiforms as well as in post-Devonian stegocephalians in which they have been termed 'meckelian foramina' or 'fenestrae' (Nilsson 1943, 1944). However, judging from conditions in *Jarvikina* (*Eusthenopteron*) *wenjukowi* (Jarvik 1937, pp. 108–109, Fig. 17; Nilsson 1944, p. 37), the exoskeletal foramina are openings of canals running in grooves on the outside of the meckelian bone. According to Nilsson (1944, Fig. 22; cf. Shishkin 1973, Figs. 51, 55, 62; Vorobyeva 1977, Figs. 16, 17) they transmitted branches of mylohyoid nerves, arteries and veins, and they may therefore most conveniently be termed *mylohyoid foramina* (*f.myl*), a name sometimes used also by Nilsson (1944, p. 36).

Of more importance are two foramina on the posterior part of the inner side of the lower jaw, one dorsal and one ventral. These foramina are well shown in A. 84 (Pl. 30:2) and of interest is that the same specimen of *Ichthyostega* also presents two openings of canals piercing the meckelian bone and exposed in the bottom of the adductor fossa (Pl.30:3). That is, the conditions in *Ichthyostega* are exactly as in *Eusthenopteron* in which, as described elsewhere (Jarvik 1980a, p. 181, Fig. 129; 1980b, pp. 167–171, Fig. 97), the dorsal canal (*f.r.md. VII*) is for the chorda tympani, whereas the ventral canal (*f.r.aut. V*) transmitted a branch of n.trigeminus known as ramus auriculotemporalis.

The postcranial skeleton

The vertebral column including its cranial portion

General remarks with a discussion of resegmentation. – In the preliminary descriptions of the vertebrae in *Eusthenopteron* and *Ichthyostega* (Jarvik 1952) it was established that the conditions in the osteolepiform fish and the early tetrapod are very similar. In both, each vertebra is composed of three paired elements and is clearly of the rhachitomous type characteristic of many post-Devonian stegocephalians. According to the current terminology these elements are the neural arch and two so-called centra, a pleurocentrum and an intercentrum. Since I found this terminology both confusing and unpractical I choose to adopt Gadow's arculia theory and has used his terminology also in the present paper in spite of Romer's categorical statement (1970, p. 158, footnote) that 'this attractive theory has (alas) proved to be untenable'.

As is well known (see e.g., Jarvik 1980b, pp. 8–9), the parachordal mesoderm early in vertebrate ontogeny is divided into a series of portions generally referred to as segments. However, since the term 'segment' is ambiguous, I have found it necessary (Jarvik 1972, pp. 229, 286) to use instead the unequivocal term *metamere*. Instead of *segmentation* we therefore have to speak of the *metamerism* of the vertebrate body. In gnathostomes, the subdivision of the parachordal mesoderm is confined to the dorsal part, adja-

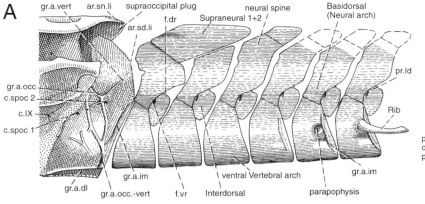

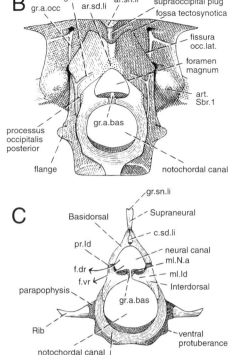

Fig. 33. Eusthenopteron foordi. □A. Restoration of posterior part of otoccipital and anterior part of vertebral column in lateral view. □B. Restoration of posterior part of otoccipital in posterior view. □C. Restoration of fifth vertebra with ribs in posterior view. After wax models in the magnification 10:1 on the basis of grinding series 2. Ribs after the prepared part of specimen P. 222; grooves on the lateral side of the occipital region after P. 6846. From Jarvik 1975, Fig. 10. *ar.sd.li, ar.sn.li,* areas for attachment of supradorsal and supraneural ligaments; *art.Sbr1,* articular area for suprapharyngobranchial 1; *c.sd.li,* canal for supradorsal ligament; *c.spoc 1, c.spoc 2,* canals for spino-occipital nerves; *c.IX,* canal for n.glossopharyngeus; *f.dr, f.vr,* foramina for dorsal and ventral roots of spinal nerves; *gr.a.bas,* groove for arteria basiralis; *gr.a.dl,* groove for lateral dorsal aorta; *gr.a.im,* groove for intermetameric artery; *gr.a.occ,* groove for occipital artery; *gr.a.occ.-vert,* groove for occipito-vertebral artery; *gr.a.vert,* groove for a.vertebralis dorsalis; *gr.sn.li,* groove for supraneural ligament; *ml.Id, ml.N.a,* medial lamellae of interdorsal and neural arch; *pr.Id,* process of interdorsal.

cent to the notochord and the spinal cord, and there arises a series of metameric portions known as somites. As is also well known, each somite becomes subdivided into three parts: the dermatome (without interest in this connection), the sclerotome forming skeleton of the vertebral column, and the myotome forming somatic musculature.

When studying the origin of the paired fins (Jarvik 1965b, pp. 154–156, Fig. 6) I found that (Jarvik 1972, p. 235) 'it is likely that the original supporting elements of the paired fins, the radials, are secondary formations arisen to form supports for pre-existing radial muscles. It is reasonable to assume that also the metameric myomeres of the trunk are older phylogenetically than the supporting elements of the vertebral column' (cf. Bjerring 1984b).

Most likely the early gnathostomes were swimming animals with a well-developed musculature. When the supporting skeletal vertebral elements, the arcualia, early in phylogeny began to develop, it is easy to understand why the alternate contractions of the metameric muscles caused a subdivision of the sclerotomic material into half-sclerotomes. Usually two such half-sclerotomes, one caudal from one metamere and one cranial from the metamere next behind, form dimetameric vertebrae separated by intrametameric vertebral joints, spanned by metameric muscles (see e.g., Jarvik 1980b, Fig. 39).

This subdivision of the sclerotomic material is generally referred to as *resegmentation,* a term still used by many (e.g., Lauder 1980; Wake & Wake 1986; Carroll 1989; Shishkin 1989). Since 'segmentation' in this case obviously bears upon

metamerism, it is in reality *remetamerism* that is claimed. It is then of great importance to note that it is only the sclerotomic material of the somite that has been subject to subdivision (segmentation). No segmentation of the myotomic material has occurred. Thus the metameric muscles have been retained, and it is the action of these unsegmented metameric muscles that has caused the subdivision of the sclerotomic material into half-sclerotomes. We are here concerned with a process that has nothing to do with the subdivision into metameres earlier in phylogeny. Under these circumstances the term 'resegmentation' (='remetamerism'), being both unneccessary and misleading, is to be rejected, and as suggested elsewhere (Jarvik 1972, p. 235) we have rather to speak of *vertebral segmentation.* With this term we thus mean the subdivision (segmentation) of the sclerotomic material into half-sclerotomes, owing to the action of the unsegmented metameric muscles, and as evidenced by analysis of the cranial vertebrae there is in this respect no difference between fish and tetrapods.

As is now well established (Jarvik 1959b, 1960, 1972, pp. 227–256; 1980a, pp. 27–28; 1980b, pp. 66–72), the vertebral column has a long cranial portion. The cranial vertebrae agree with those in the trunk and are also in the head intermetameric in position and composed of material derived from half-sclerotomes of two adjoining somites. The four paired arcual elements (Gadow's arcualia), two anterior (basidorsal and basiventral) derived from the caudal half-sclerotome of one metamere and two posterior (interdorsal and interventral) derived from the cranial half-sclerotome of the metamere following next behind, are well known in many fishes (e.g., in *Amia*; Jarvik 1980a, Schultze & Arratia 1986). With the aid of the metameric basicranial muscles, which are described by Regel (1961, 1964, 1968) and Bjerring (1970) and well known in particular in larval urodeles, and of other embryological data, it has been possible to identify these elements also in the cranial vertebrae (Jarvik 1972, pp. 241–256; 1980b, pp. 66–72). It could be demonstrated that the interdorsal in urodeles has fused with the basidorsal to form a complex neural arch, as is the case also in extant anurans, reptiles and birds (Smit 1953, p. 121; as for reptiles see also Jarvik 1980b, pp. 154–156). In these cases not only the basidorsal but also the interdorsal obviously are provided with dorsal ascending processes, which together form the complex neural arch (Jarvik 1972, p. 245; 1980b, p. 65). Also in urodeles, the basiventral has fused with the interventral, forming a complex structure which I since 1952, in order to avoid the abstruse term 'intercentrum' (cf.p. 48), has termed the *ventral vertebral arch*. This structure has been recognized also in reptiles, in which it forms the so-called primary centre (Jarvik 1980b, Fig. 89). The rejection of Gadow's arcual theory by Romer and others was obviously too rash. As a matter of fact, it is only on the basis of this theory that it has been possible to unravel the cranial vertebrae (to speak of pleuro- and intercentra in these vertebrae is certainly not very attractive). After these general considerations we may first turn to the conditions in *Eusthenopteron*.

EUSTHENOPTERON. – My first description of the vertebral column in *Eusthenopteron*, dealing mainly with its posterior part (Jarvik 1952), was based on specimen P. 222 and other specimens prepared mechanically. Later (Jarvik 1975), the anterior part was described on the basis of the grinding series and wax models of the well-preserved skull of P. 222, and striking similarities with the adjoining part of the neural endocranium, formed by fused cranial vertebrae, were demonstrated (Fig. 33).

A series of supraneurals continued in the endocranium by the supraoccipital plug, was discovered. The posterior tip of the plug situated between the openings of the canals for the occipital arteries lacks periosteal lining and forms the attachment area for the supraneural ligament *(ar.sn.li)* which ran backwards in a groove at the dorsal ends of the supraneurals. Another area devoid of periosteal lining *(ar.sd.li)* situated immediately dorsal to the foramen magnum is obviously the area of attachment for the supradorsal ligament which in the vertebrae runs in a separate canal *(c.sd.li,* Jarvik 1980a, Fig 13; 1980b, Fig. 39). Moreover, it was shown that the paired basidorsals and interdorsals are provided with medial laminae that together form a horizontal wall between the canals for the medulla spinalis and the notochord (Fig. 33C). This wall is continued in the posterior part of the otoccipital, and as in the vertebrae there is a median groove *(gr.a.bas)* for the basilar artery on its dorsal side (Fig. 33B).

Lateral to the horizontal wall the part of the otoccipital, obviously formed by a posterior cranial interdorsal, projects into a posterior process also lacking periosteal lining on its top. Since, according to de Beer (1937, p. 385), the *occipital condyle* in extant amphibians is derived from a posterior cranial interdorsal, the paired posterior occipital process in *Eusthenopteron* (Fig. 33B) 'may be regarded as an occipital condyle in an initial stage of development' (Jarvik 1975, p. 204).

The ventral vertebral arch is, as is readily seen, in the posterior part of the otoccipital represented by the portion that forms the main part of the walls of the canal for the notochord. It is then of interest to note that this part on the outside presents a groove for an intermetameric artery *(gr. a. im,* Fig. 33A)* as on most of the independent ventral vertebral arches in the vertebral column. In the anterior part of the otoccipital there may be an independent subchordal (ventral) arcual plate (zygal plate) probably representing a ventral vertebral arch. This plate is imperfectly preserved in *Eusthenopteron,* but judging from the configuration of the neurocranial base and conditions in porolepiforms and extant anurans (Jarvik 1972, pp. 65–69, Figs. 22, 24G) it was an arched structure similar to the ventral vertebral arches in the vertebrae (Bjerring 1973, Fig. 7; Jarvik 1980a, Fig. 92). Sometimes the ventral arcual plate is incorporated in the otoccipital, in which case the basicranial fenestra is relatively short (Jarvik 1954, Fig. 7; 1972, Fig. 93E; 1980a, Fig. 93B). In *Spodichthys* (Jarvik 1985, pp. 39–40, Figs. 24B, 26A), a subchordal arcual plate is shown in several specimens in front of the fissura oticalis ventralis anterior. Since this plate is relatively short it was suggested (Jarvik 1985, p. 40) that it 'represents only the part of CV 3+4 belonging to the fourth metamere, that is interventral 4'.

The horizontal wall in the posterior part of the otoccipital, separating the spaces for the brain and the notochord, is continued forwards by two suprachordal arcual plates (Jarvik 1980a, Fig. 91B; anazygals 2p, 3a, Bjerring 1971, Fig. 2B) extending forwards to or almost to the suprachordal horizontal wall at the posterior end of the ethmosphenoid.

It was the vertebra-like appearance of the posterior end of the ethmosphenoid (Jarvik 1972, p. 232, Fig. 93A–C; 1980b, Fig. 38C) that caused me to suggest (Jarvik 1959b, pp. 63, 69; 1960. pp. 19, 87) that the vertebral column has a cranial portion. Since this end obviously is formed by a cranial vertebra it is readily seen that the intercranial joint is a persisting vertebral joint, and as is now well known, this joint was spanned by a metameric basicranial muscle (the subcranial muscle or basicranial muscle 3).

Farther to the rear, behind the basicranial fenestra, the vertebral joints are represented by fissurae in the basis cranii (fissura oticalis ventralis anterior, fissura oticalis ventralis posterior) or in the side of the endocranium (fissura occipitalis lateralis). As evidenced by distinct impressions also these vertebral joints were spanned by metameric basicranial muscles (Jarvik 1972, Fig. 99; 1980a, Fig. 93B; 1980b, Fig. 40).

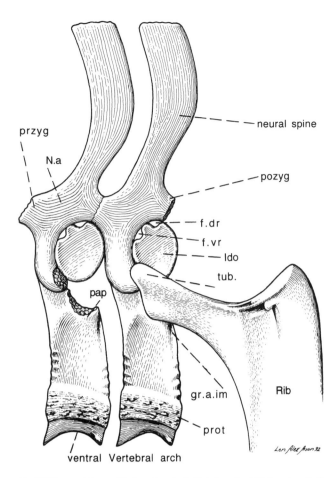

Fig. 34. Two vertebrae with dorsal part of adjoining rib of *Ichthyostega*. Left side in lateral aspect. *Ido*, interdorsal; *N.a*, neural arch; *f.dr, f.vr*, foramina for dorsal and ventral roots of spinal nerves; *gr.a.im*, groove for intermetameric artery; *pap*, parapophysis; *pozyg*, postzygapophysis; *przyg*, prezygapophysis; *prot*, ventral protuberance; *tub*, tuberculum.

Ichthyostega. – The vertebrae in *Ichthyostega* (Figs. 34, 38B, 47, Pls. 34–40) are as in *Eusthenopteron* (Figs. 33, 38) composed of three paired elements; the neural arch (basidorsal), the interdorsal and the complex ventral vertebral arch (basiventral and interventral). The interdorsal also in *Ichthyostega* (Pl. 34) shows notches for the passages of the dorsal (*f.dr*) and ventral (*f.vr*) roots of the spinal nerves. The neural arches are provided with distinct pre- and postzygapophyses, and there is a ventral articular area for the tuberculum of the rib. The articular area for the capitulum is found on a distinct parapophysis (*pap*) on the outside of the ventral vertebral arch, and close posteroventrally to the parapophysis is a short groove (*gr.a.im*) for the intermetameric artery (A.251, Pl.35:6). A ventral protuberance (*prot*. Fig. 34; Pl.35:2, 6) similar to that in *Eusthenopteron* (Fig. 33C; Jarvik 1980a, Fig. 96) is present, as is also a canal for the supradorsal ligament (*c. sd.li*, Pls. 37, 38). Most likely, this ligament was attached to the parietoextrascapular (*ar. sd.li*, Fig. 35) immediately dorsal to the notch which represents the foramen magnum. Dorsal to this area of attachment, the posterior end of the parieto-extrascapular, between the openings of the canals for the occipital arteries (*ar.sn.li*, Pl.2:1, 3), presents a median depression (*ar.sn.li*, Pl.2:1, 3), which, as pointed out above (p. 50), evidently is the area of attachment for the supraneural ligament (Figs. 35, 36). In these respects, *Ichthyostega* agrees with certain extant anurans (*Pelobates fuscus, P. cultripres* Jarvik 1975, Fig. 11), in which, in contrast to conditions in *Eusthenopteron* (Fig. 33B), both ligaments are attached to the posterior end of the exoskeletal skull roof. It is also to be noticed that the openings of the canals for the occipital arteries (*c.a.occ*) in *Ichthyostega* lie close together, somewhat as in *Pelobates cultripres*.

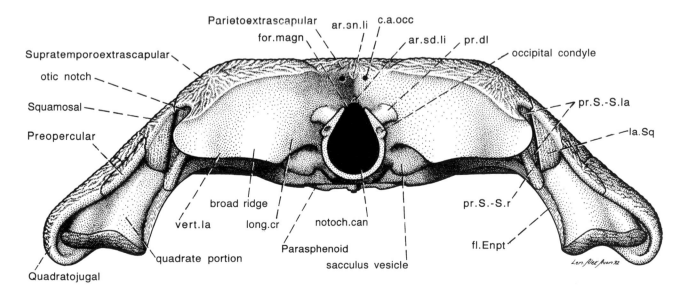

Fig. 35. Restoration of skull of *Ichthyostega* in posterior view. *ar.sd.li, ar.sn.li*, areas of attachment for supradorsal and supraneural ligaments; *c.a.occ*, canal for occipital artery; *fl.Enpt*, flange of quadrate ramus of entopterygoid; *for.magn*, foramen magnum; *la.Sq*, squamosal lamina; *long.cr*, longitudinal crest; *notoch.can*, canal for the notochord; *pr.dl*, dorsolateral process; *pr.S.-S.la, pr.S.-S.r*, laminar and rodlike portions of Säve-Söderbergh's ventral process; *vert.la*, exoskeletal part of complex vertical lamina.

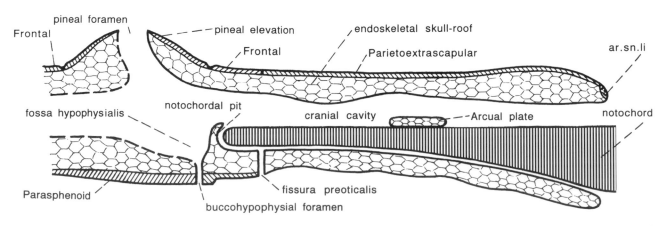

Fig. 36. Restoration of median longitudinal section of posterior part of skull of *Ichthyostega* to show the fissura preoticalis and anterior extent of notochord. *ar.sn.li*, attachment area for supraneural ligament.

It is remarkable that *Ichthyostega* lacks a distinct horizontal wall separating the spaces for the neural tube and the notochord in the present state of preservation is lacking both in the vertebrae (Pl. 35:1, 4, 5; Jarvik 1980a, Fig. 157A) and in the posterior part of the otoccipital (Fig. 35). However, since this wall is present in *Eusthenopteron* and may be represented by a membrane as in *Amia* (Jarvik 1980a, Fig. 13), by a suprachordal arcual plate as in *Latimeria*, or by suprachordal cartilages as in extant anurans (Jarvik 1972, p. 65, Fig. 24), it is likely that some kind of wall was present in ichthyostegids. It is possible that the plates in the roof of the cranial cavity in one specimen of *Ichthyostega* (Pl. 25:2, 3) are displaced suprachordal arcual plates, and in some other specimens (e.g., A. 64, A. 152, A. 251) there are skeletal pieces that may be remains of the horizontal wall. However, in the two specimens cut longitudinally (Pls. 16:3, 4; 21:1, 2), such a wall is lacking most posteriorly, whereas farther forwards in the otoccipital a small endoskeletal plate is present in both specimens. Judging from its position, this plate separates the spaces for the brain and the notochord and is a dorsal (suprachordal) *arcual plate* (see Jarvik 1980a, pp. 127–130, Figs. 91, 92; anazygal, Bjerring 1971, p. 191). As shown in one of the specimens (Pl. 21:2), the space for the notochord extends forwards to a small pit in the posterior wall of a cavity in the rear of the orbitotemporal region. This cavity is continued ventrally by the buccohypophysial canal piercing the parasphenoid, and therefore it must be the fossa hypophysialis.

These conditions indicate that the notochord in *Ichthyostega*, as tentatively shown in Fig. 36, in contrast to conditions in *Eusthenopteron* and other 'crossopterygians' (Jarvik 1980a, Figs. 91, 214), tapered in thickness forwards. It passed dorsal to the fissura preoticalis, which is a vestige of the osteolepiform intracranial joint, and, as in 'crossopterygians' in general, it reached the posterior end of the ethmo-sphenoid. There its anterior end was received by a *notochordal pit* in the same position as, but smaller, than the

corresponding pit in *Eusthenopteron* and other 'crossopterygians'. In that group the basicranial fenestra lies behind this pit; in *Ichthyostega* no such opening is present. The ventral wall of the canal for the notochord extends forwards to the fissura preoticalis (Fig. 36), a condition that no doubt is due to the incorporation of subchordal arcual plates (catazygals) such as found separate in many 'crossopterygians' (see, e.g., Bjerring 1967, 1971, 1973, 1977, 1978; Jarvik 1972, pp. 65–69; 1980a, pp. 127–130; 1985, pp. 39–40). Close behind the fissura preoticalis, the external side of the ventral wall of the canal for the notochord in *Ichthyostega* presents the depression that has been interpreted (p. 45) as the posterior area of attachment for the paired basicranial muscle 3 (subcranial muscle), which spanned the vertebral joint represented by the fissura (*bcr.mu*, Fig. 37, Pls. 4:3; 16:1; 26:2; 27:2; 28). Behind this depression, as recorded earlier (Jarvik 1980a, p. 236, Figs. 171, 172), the external side of the ventral wall of the canal for the notochord in A.64 (Pl. 4:3) shows four elevations separated by arched transverse grooves (Fig. 37). The elevations represent the ventral vertebral arches of four cranial vertebrae (CV) which, according to my interpretations (Jarvik 1980a, pp. 68–72, Figs. 40–42), are CV 3+4, CV 4+5, CV 5+6 and CV 6+7. The grooves between the elevations are vestigial fissurae (fissura oticalis ventralis anterior, fissura oticalis ventralis posterior, and fissura occipitalis) marking the position of former vertebral joints. Such arched transverse grooves are found also in other specimens, and a remarkable fact is the presence of peculiar striated processes (*str. pr*) at the ends of the grooves (Fig. 37; Pls. 21:4; 24:4; 26:2; 27:2; 28). The importance of these processes is unknown, but it may be noted that the infrapharyngeals of the first and second branchial arches in *Eusthenopteron* (Jarvik 1980a, Figs. 86, 124) articulate with the neural endocranium in corresponding positions. In this connection it may also be noted that impressions for the basicranial muscles spanning the vertebral joints, corresponding to those found in *Eusthenopteron* (Jarvik 1980a, Fig 40), are lacking in *Ichthyostega*. It

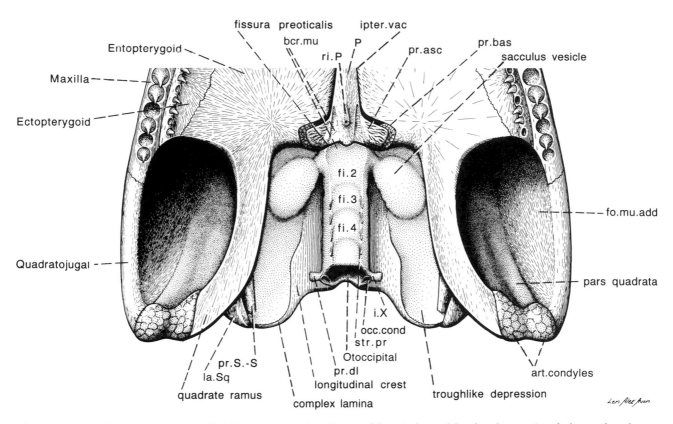

Fig. 37. Restoration of posterior part of palate of *Ichthyostega*. *P*, parasphenoid; *art.condyles*, articular condyles of quadrate portion of palatoquadrate; *bcr.mu*, anterior and posterior areas of attachment for basicranial muscles; *fi.2*, fissura oticalis ventralis anterior; *fi.3*, fissura oticalis ventralis posterior; *fi.4*, fissura occipitalis; *fo.mu.add*, fossa for adductor muscle; *ipter.vac*, interpterygoid vacuity; *i.X*, notch probably for n.vagus; *la.Sq*, squamosal lamina; *occ.cond*, occipital condyle; *pr.asc*, processus ascendens of parasphenoid; *pr.bas*, processus basalis; *pr.dl*, dorsolateral process of otoccipital; *pr.S.-S*, the process of Säve-Söderbergh; *ri.P*, median ridge of parasphenoid; *str.pr*, striated processes of otoccipital.

may be added that a median ridge is shown on the external side of the ventral wall of the canal for the notochord in one specimen (Pls. 20:1; 26:2), but there is no groove for the lateral dorsal aorta such as found in osteolepiforms.

At the posterior end of the otoccipital in *Ichthyostega* (Figs. 35, 37), there are two paired processes, one posterior and one dorsolateral. The posterior process (Pls. 22:1; 25:2, 3), which has been interpreted as the occipital condyle (Jarvik 1952, Fig. 36), is situated much as that condyle in *Eusthenopteron* (processus occipitalis posterior: Fig. 33B) and in extant anurans (Jarvik 1975, Fig. 11). The dorsolateral process (*pr.dl*, Pls. 25:3; 26:1) meets a process on the medial (exoskeletal) side of the longitudinal crest in the palate (Fig. 37). The concavity (*i.X*) on the ventral side of the structure thus formed may perhaps be for the passage of the vagus nerve.

The fish-like tail

Material. – The tail in *Ichthyostega* is provided with a median fin, which, as in fishes but in contrast to post-Devonian tetrapods, is supported by endoskeletal fin-supports, radials, and dermal fin-rays (lepidotrichia). The description (Jarvik 1952) was based on four specimens: A.69 (Pl. 36), Smith

Woodward Bjerg, Coll. 1936; A.109 (Pls. 37, 38, 39:1), Celsius Bjerg, Coll. 1948; A.156 (Pl. 40) and A.157 (Pl. 39:2) Sederholm Bjerg, 1174 m, Coll.1949). Because no new material has been added, only a summary of my previous description will be given. (As to the scales shown in A.109, see 1952, pp. 46–47.)

Vertebral column and median fin-supports. – The vertebral column (Figs. 38B, 47; Pls. 36–40), from the sacral region backwards, first forms a gentle curve and then continues to the tip of the tail, where it turns slightly upwards. In the posterior part of the tail the vertebrae have fused into a bony rod, the urostyle, pierced by a narrow canal (*c.not*, Pl. 39:1) for the posterior end of the notochord. In front of that rod the vertebral column is poorly ossified in its central part and is represented mainly by the neural and haemal spines and the adjoining parts of the neural and ventral vertebral arches. Interdorsals are present but are imperfectly preserved. As in *Eusthenopteron* (Fig. 33), the paired neural arches enclose canals for the spinal cord and the supradorsal ligament (Pls. 37, 38). The haemal canal is also present (Pls. 38:2, 39:1, 40). In the main posterior part of the tail the neural spines, decreasing in height, are inclined backwards and articulate

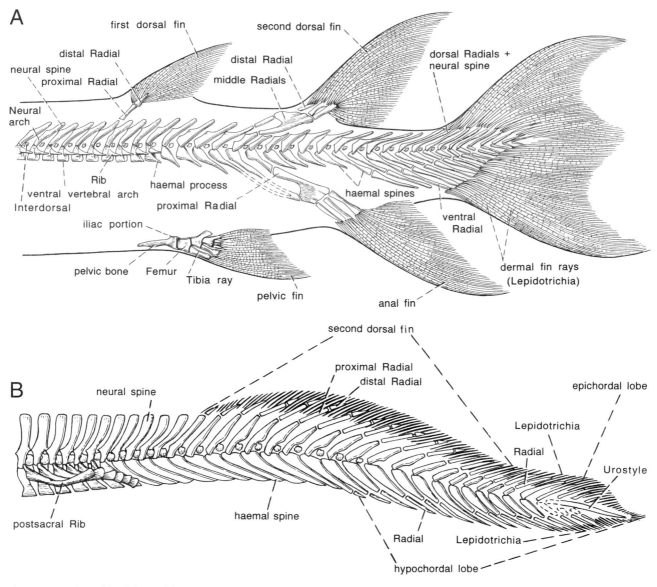

Fig. 38. Restorations of the skeleton of the tail in (A) *Eusthenopteron* and (B) *Ichthyostega* in lateral aspect. From Jarvik 1952.

with a corresponding number of radials. The haemal spines articulate with radials only in the posterior half of the tail. In the anterior half, where ventral radials are lacking, they are comparatively stout and about equal in length. Moreover, they lie close together and are posteriorly directed, conditions which cause a distinct shallow depression in the ventral margin of the tail (Fig. 38B).

The endoskeletal fin-supports, the radials, are long and slender rods pointed distally and arranged in two series, dorsal and ventral. As discovered later (Jarvik 1959a, p. 36), at least some of the radials in the dorsal series are divided into a long proximal and a short distal element (*Rp, Rd* Pl. 39:2). These subdivided radials are found in the anterior portion of the continuous dorsal fin, more precisely in the portion that has been assumed to correspond to the second dorsal fin in the osteolepiforms. This is of some interest,

since the radials in the second dorsal fin in osteolepiforms, at least in *Eusthenopteron* (Fig. 38), are also divided into proximal and distal elements. Moreover, the radials in the epichordal lobe of the caudal fin in *Ichthyostega* have fused with the neural spines, whereas in the hypochordal lobe they are independent. In this respect, too, the ichthyostegids agree with *Eusthenopteron*.

Of particular importance is that the dorsal series of radials extends farther forwards than the ventral one and includes about 24 elements, against about 16 in the ventral series. The dorsal radials run in the same directions as the neural spines, most of them articulating with the distal ends of the latter. All the dorsal radials support dermal fin-rays, whereas in the ventral series only the hindmost radials are connected with such rays. These hindmost ventral radials resemble the dorsal radials and articulate with the distal ends of the haemal spines, whereas the anterior ventral radials, about ten in number, are slightly modified. These elements, which decrease in size forwards, articulate with the distal ends of the haemal spines, too. However, they are slightly bent backwards and lie comparatively close together. These radials form the most ventral part of the skeleton of the tail, and no doubt the modifications and the absence of dermal fin-rays (see below) are connected with the fact that this part of the tail dragged on the ground when the animal was walking in shallow water or on land.

Dermal fin-rays and the median fin. – The dermal fin-rays in *Ichthyostega* (Pls. 36–40), supported by the radials and most posteriorly by the urostyle, are more numerous than the radials, as in osteolepiforms and most other fishes. As is characteristic for many early fishes, the fin-rays contain distinct spaces for bone cells (Jarvik 1952, Fig. 4; 1959a, pp. 31, 37). Most often they are of about the same length as the radials, nearly round as seen in transverse sections and tapering towards the pointed distal end. However, they never become so thin towards the periphery as in many fishes, and they are most suggestive of the comparatively long proximal segments of the fin-rays in osteolepiforms and porolepiforms (Jarvik 1959a, Figs. 9, 10). Like these segments they are, as far as can be seen, unjointed, and their pointed proximal ends embrace the distal parts of the radials. However, whereas in the osteolepiforms (Jarvik 1959a, Fig. 10A, B) the fin-rays overlap only the distal ends of the radials, in *Ichthyostega* they overlap the radials to a considerable extent, and generally only about one third of the rays extending beyond the distal tips of the radials. However, in the dorsal shallow depression of the median fin, well shown in particular in A.157 (Pl. 39:2), the fin-rays are comparatively short, hardly reaching beyond the distal ends of the radials.

As evidenced by these and other data recorded above, it is likely that the dorsal depression in the continuous median fin in ichthyostegids marks the boundary between two incompletely separated fins: one fairly long dorsal fin, corresponding to the second dorsal fin in osteolepiforms, and a caudal fin. As the heterocercal caudal fin in early osteolepiforms (*Osteolepis, Thursius*, Jarvik 1948a, Figs. 26, 27, 57, 64, 70), that of *Ichthyostega* consists of a short epichordal lobe and a longer hypochordal lobe, which may be regarded as ending anteriorly at the most anterior of the ventral radials.

The absence of fin-rays in the anterior part of the hypochordal lobe in ichthyostegids and the modifications of the radials in this region of the tail – that is, the region that, as explained above, obviously rested on the ground and was dragged in the mud when the animal was walking in shallow water or on land – renders it likely that the fin-rays in this part were reduced and disappeared in connection with the change from water life to an amphibious mode of life. Also, the thin distal end parts of the fin-rays were reduced at the same time. The retention of fin-rays in the most posterior part of the hypochordal lobe is obviously due to the fact the posterior part of the tail is slightly upturned.

Ribs

In the outstanding specimen A.115, the posterior part of the skull is preserved (Pl. 21:5). Behind that follows a considerable part of the rib cage (Pl.42). The stout and elongated ribs, about ten in number on each side, form a rigid armour encircling the body cavity. Ventrally the ribs nearly meet in the median line (Pl. 42:3) but no traces of sternal structures have been found. The structure (Figs. 34, 39) is best shown on isolated ribs (Pls. 43, 44; 50:1) and several latex casts of such ribs have been made. When the counterpart is preserved and the skeletal tissue has been removed, it is possible to make a solid latex cast showing the structure of the rib as whole (Pl. 44:1, 2).

Each of the long and robust thoracic ribs includes an anterior arched rodlike and a posterior laminar portion. Proximally the two portions together form a strong shaft, the rodlike portion terminating at the slightly expanded head with the capitulum and the laminar portion with the tuberculum. In two respects, *Ichthyostega* differs from post-Devo-

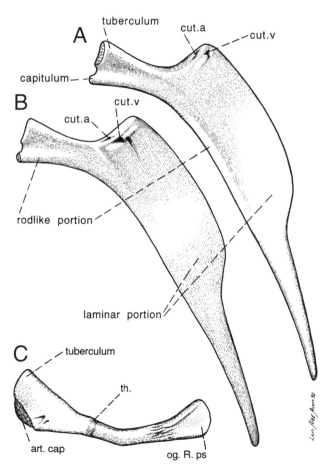

Fig. 39. Restorations of ribs in *Ichthyostega.* □A, B. Thoracal ribs in external (lateral) and internal (medial) views. □C. Postsacral rib in internal view. *art.cap*, articular area of capitulum; *cut.a, cut.v*, canals for cutaneous artery and vein; *og.R.ps*, area overlapping postsacral rib following next behind; *th*, thickening.

nian stegocephalians. As shown in specimens 99 (Pl. 41; Stensiö 1931, Pl. 36) and A. 115 (Pl. 42), the broad laminar portion overlaps the succeeding rib. Moreover, whereas ribs in post-Devonian stegocephalians rarely may be pierced by a small foramen for a nerve or blood vessel (Warren & Snell 1991, p. 51), a characteristic feature of *Ichthyostega* is that the dorsal part of the laminar portion is pierced by two canals – one dorsal, rather narrow (*cut.a*) and one much wider ventral (*cut.v* Fig. 39; Pls. 41:2, 43, 44). These canals, concealed from the outside when the ribs are in their natural position, form two longitudinal passages, one dorsal, conceivably for an artery comparable to the arteria cutanea magna, and one ventral, conceivably for a vein comparable to the vena cutanea magna in the recent frog *Rana* (see Gaupp 1899, Fig. 84).

There is, accordingly, evidence suggesting that, as in the frog, branches were given off from the cutaneous blood vessels. A rather broad groove for such a branch of a vein is seen in A.235 (*gr. cut* Pl. 44:3) and in A.173, (Pl. 44:1, 2) small foramina (*f. cut*), obviously for arterial branches, are discernible. The presence of longitudinal passages for blood vessels together with the fact that at least the main part of the body was naked (small cycloid scales have been found only on the tail) renders it likely that the ichthyostegids, at least partly, were skin breathers, as in frogs both in water and on land (Gaupp 1904, pp. 461–463).

Among the ribs shown, A.36 (Pl. 43:5) is characterized by an elongated shaft. Most likely, this rib and possibly also A.25 (Pl. 43:4) belong to the lumbar region. Moreover, postsacral (caudal) ribs are shown in A.109 (Pls. 37, 38), A.54, and A.140 (Pl. 34:1, 2). In the latter two specimens, the ribs are directed backwards, as in *Eogyrinus* (Panchen 1972, Fig. 16) and *Archeria* (Holmes 1989, Fig. 16). It is of interest also that they run parallel with and are suggestive of the posterior iliac process of the pelvic bone 'which thus presumably represents a postsacral rib secondarily incorporated in the pelvis' (Jarvik 1952, p. 39). This is well shown on latex casts of A.54 made after 1952, showing also the iliac process of one of the pelvic bones. It is of interest also that these casts present an isolated postsacral rib in front of the five ribs shown in my restoration of the tail on the basis of a plastic cast (Jarvik 1952, Fig. 14A). This rib lies close inside the iliac process, and the external side is therefore concealed. However, the internal side is well shown (Pl. 43:6). The gently arched dorsal margin of the rib (Fig. 39C) follows exactly the curvature of the dorsal margin of the iliac process situated close outside the rib and is of the same length. The proximal (anterior) end of the bone is expanded and presents, in addition to the articular areas of the tuberculum and capitulum, two grooves leading to small foramina (*cf. cut.a, cut. v,* Fig. 39B). The middle part, which shows a thickening (*th*), is rodlike, and there is no laminar portion. Backwards, the middle part merges into the partly ornamented distal (posterior) part, which ventrally presents a distinctly depressed area (*og.R.ps*) overlapping the succeeding postsacral rib (Figs. 38B, 47).

Appendicular skeleton

General remarks. – As stated by Jarvik (1964, pp. 76–77) 'the fundamental structures characteristic of both the girdles and the limbs of the early tetrapods arose already in their piscine ancestors. Since a typical eutetrapod limb occurred, in a practically final stage, within the paired paddles of the osteolepiforms it is evident that the transformations at the transition from fish to tetrapod were inconsiderable.' Among the changes may first be noticed the progressive development of the endoskeletal girdles accompanied by a partial reduction of the exoskeletal shoulder girdle. Comparing *Eusthenopteron* with *Ichthyostega* and post-Devonian early tetrapods, it is evident that the pelvic girdle has grown ventrally to meet its antimere in a median symphysis, whereas the iliac portion has grown upwards to gain contact with one or more sacral ribs.

The endoskeletal shoulder girdle in *Ichthyostega* (Jarvik 1964, p. 73) 'resembles and is easily derivable from that of the osteolepiforms'. It includes a glenoid portion corresponding to the entire girdle in *Eusthenopteron* but, as in the pelvic girdle, it has grown downwards and a strong coracoid plate has been added. However, no upward growth is discernible, and since the scapular blade is lacking, the current term 'scapulocoracoid' is inappropriate (cf. *Eusthenopteron*, Jarvik 1980a, p. 139).

Shoulder girdle and forelimb. – A characteristic feature of the shoulder girdle in *Ichthyostega* is that the cleithrum has fused with the endoskeletal shoulder girdle. The solid composite structure thus formed is found isolated in many specimens (Pls. 47–49, 50:1); it is present on both sides together with the posterior part of the skull and the rib cage in A.115 (Pls. 45, 46). As seen on the left side of that specimen the dorsal end of the cleithrum (Pl. 42:1) is situated outside the rib cage and rather far behind the skull roof, which in *Ichthyostega* includes the extrascapular series. In *Eusthenopteron*, a row of three bones (posttemporal, supracleithrum and anocleithrum) connects the extrascapulars with the dorsal end of the cleithrum, but it is not known if such bones are present in *Ichthyostega*.

Exoskeletal shoulder girdle. – The cleithrum in *Ichthyostega* (Figs. 40, 42; Pls. 45, 49, 50:1) differs considerably from that in post-Devonian stegocephalians (Nilsson 1939, pp. 16–24; Romer 1947, 1957; Carroll 1967; Brough & Brough 1967; Holmes 1980; Panchen 1985; Clack 1987; Godfrey 1989). It is more completely developed than in these forms and may be compared with the dorsal rectangular portion of the cleithrum in *Eusthenopteron*. Like the latter, it is provided with an anteroventral process (Pls. 45:1, 47:4) but in contrast to *Eusthenopteron* its ventral part, intimately joined to the endoskeletal girdle, is considerably thickened and forms on the inner side a pronounced knob, the *medial elevation* (Fig. 42A). This elevation (Pls. 45:2–4, 50:1) is situated dorsal to a depression, the *clavicular recess* (A.115, Pl. 45:2, A.182, Pl.

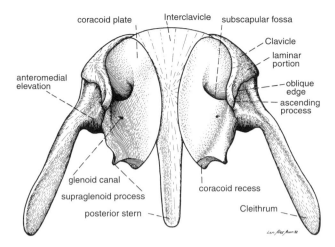

Fig. 40. Restoration of shoulder girdles of *Ichthyostega* in dorsal aspect.

50:1), occupied by the dorsal end of the ascending process of the clavicle. Dorsally the medial elevation is continued by a medial ridge tapering towards the dorsal end of the cleithrum. In front of that ridge there is an anteromedial depression along the anterior margin of the bone.

On the anterior side of the medial elevation, the cleithrum continues downwards to the tip of the anteroventral cleithral process as a distinct *ventromedial lamina*. This lamina (Pls. 45: 2, 5; 47:1), which forms the broadened posterior wall of the anteromedial depression, has a free margin, the *oblique edge* (Pls. 45:5; 47:1; 50:1). As is well shown on a latex cast of A.182 (Pl. 50:1) in which the ventromedial lamina is broken, this lamina is supported by a prominent coracoid crest formed by the anterodorsal part of the endoskeletal girdle, which is bent inwards (Fig. 42A). This crest continues upwards along the oblique edge to the clavicular recess where, turning backwards, it is exposed and then merges into the dorsal end of the supraglenoid buttress. The clavicular recess is bordered anteriorly by the dorsal part of the oblique edge. In A.182 (Pl. 50:1) this edge presents ridges and grooves that fit into corresponding ridges and grooves that exist close to the dorsal end of the rodlike portion of the clavicular ascending process.

In the present state of preservation, the external surface of the cleithrum is striate. However, in A.115 (Pl. 51:1) the ventral part of that surface, in front of a distinct depression (*depr* Pls. 45:1; 47:1), bears an ornamentation of faint ridges; judging from this specimen it is likely that at least the main part of the surface was ornamented in a similar way.

The *clavicle* (Figs. 40, 41; Pls. 45, 50, 51:2–7) includes a ventral plate, which laterally bends upwards and is continued by an ascending process. The anteromedial part of the ventral plate, which overlaps the interclavicle, is striate on the inner side (Pl. 50:1), as is the corresponding overlapped area of the interclavicle. The external surface may be ornamented with low ridges (Pl. 50:2). The ascending process extends

upwards on the inner side of the cleithrum, as it does in *Eusthenopteron,* but is more intricate in structure than in the latter. The slightly twisted dorsal part that occupies the cleithral clavicular recess (Pl. 45:2) consists, as in *Greererpeton* (Godfrey 1989, p. 101), of a rodlike anteromedial and a laminar posterolateral portion. The rodlike portion ends dorsally in the clavicular recess on the ventral side of the cleithral medial elevation and, as mentioned above, its dorsal part is provided with ridges and grooves (cf. *Greererpeton*) that fit into corresponding structures on the dorsal part of the cleithral oblique edge (Pl. 50:1). Ventral to these grooved parts, the free margin of the laminar portion abuts against the oblique edge; since the free margin is striate, as is well shown in A.189 (Pl. 50:1), it is likely that the oblique edge is striated in a corresponding way.

The *interclavicle* in *Ichthyostega* (Figs. 40, 41A) is more or less completely preserved in four specimens (Pl. 52). It consists of an anterior broad plate and a fairly long posterior stem. As is well shown in A.39 (Pl. 52:1), the ventral side of the broad plate presents a median ridge separating the large anterolateral areas overlapped by the clavicles. As mentioned above, these areas are striate, whereas the median ridge and the adjoining anterior part of the stem bears an ornamentation. The anteriorly rounded plate of *Ichthyostega*, with the long stem, differs from the more or less diamond-shaped interclavicle in most post-Devonian stegocephalians (Nils-

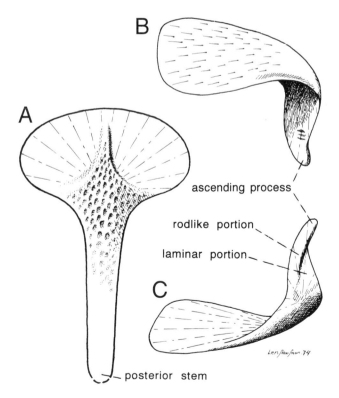

Fig. 41. Restorations of (A) interclavicle of *Ichthyostega* in ventral and (B, C) clavicle in ventral and lateral aspect.

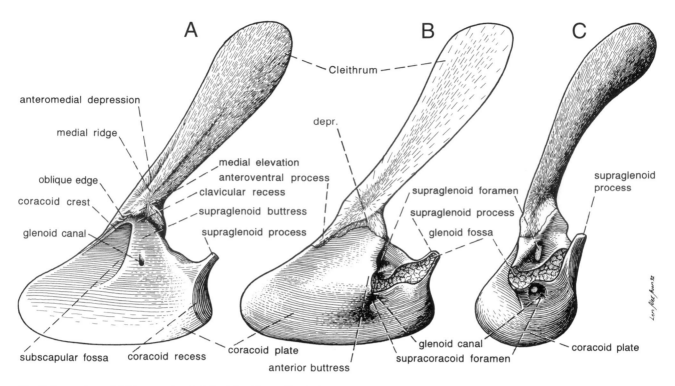

Fig. 42. Restorations of cleithrum and endoskeletal shoulder girdle of *Ichthyostega* in (A) medial, (B) lateral and (C) posterior aspects. *depr,* depression on external side of cleithrum.

son 1939, Fig. 4; Romer 1947, Figs. 13, 14; 1957, Milner 1980, Holmes 1980, Clack 1987, Godfrey 1989) but agrees with *Gephyrostegus* (Brough & Brough 1967) and certain other anthracosaurs (Romer 1947, Fig. 14, Carroll 1967).

Endoskeletal shoulder girdle. – This girdle ('scapulocoracoid') in *Ichthyostega* (Figs. 40, 42) is a solid structure, more or less well preserved in several specimens (Pls. 45–49, 50:1). It agrees in most regards with that found, e.g., in *Eryops* (Miner 1925) and *Archeria* (Romer 1957). The fact that no distinct scapular blade is discernible appears to be a consequence of the intimate connection with the cleithrum, and it is in fact difficult to decide exactly where the endoskeletal bone ends dorsally in the region of the medial cleithral elevation.

The smooth anterior part of the external side of the girdle is convex, obviously following the curvature of the inner side of the reduced ventral triangular portion of the cleithrum, as in the ancestral fish. Behind that smooth area follows the *glenoid fossa* (Fig. 42, B, C), which has been freed from matrix in several specimens (Pls. 45:1, 4; 46:1, 4, 5; 47:1, 2, 4; 48:1). In A. 115 (Pls. 45, 46), this fossa is well shown on both the left and the right girdle. On the right girdle it is situated close to the caput humeri. On both girdles it is bounded anteriorly by an elevation, probably corresponding to the anterior buttress in *Archeria* (Romer 1957, p. 108). On the left girdle it extends to the external side of the supraglenoid process, a condition which I overlooked when I prepared the restoration presented in 1980a (Fig. 165D). Immediately dorsal and ventral to the constricted middle part of the glenoid fossa, the

supraglenoid and *supracoracoid foramina,* both leading into the bottom of the deep *subscapular fossa* (Pl. 49:1, 3, 5), are more or less well shown in several specimens (Pls. 45:4; 46:1, 4;47:2, 4; 48:1). Close behind the supracoracoid foramen lies the ventral opening of the *glenoid canal* (A.249, Pl. 47:4).

The dorsal opening of the glenoid canal is found on the inner side of the girdle a little behind the ventral part of the opening of the subscapular fossa (Pls. 45:2, 3; 47:3; 48:2, 3; 50:1). In one specimen (Pl. 49:1, 2) there are two openings in the corresponding place. Anterodorsally the marginal part of the endoskeletal girdle is bent inwards and forms a prominent *coracoid crest* (Pl. 50:1) which, as described above, dorsally merges into the dorsal end of the supraglenoid buttress. Another characteristic feature of the inner side of the girdle is the crescent-shaped *coracoid recess* (Fig. 42A) found at the posterior margin and extending upwards to the supraglenoid process when this process is present (Pl. 45:2–4). The importance of this recess is unknown, but it may be tentatively suggested that it together with its antimere lodged the anterior part of the pericardium.

Forelimb. – Skeletal elements of the forelimb of *Ichthyostega* (Figs, 44–46) worth consideration are found only in one specimen in our collection. In this specimen (A. 115, Pls. 46, 53, 54), which shows also the shoulder girdles described above, the left humerus together with the radius are preserved but have been displaced backwards and were found in the cavity of the rib cage (see Pl. 42:3). However, after a difficult preparation they have been freed from the embedding hard red sandstone and have been taken out from the rib cage (Fig. 44; Pl. 53). The well-ossified and perfectly preserved humerus can therefore be studied in all orientations.

During preparation the left humerus was sawn through, but the gap made by the saw cut was replaced with plaster (Fig. 44A; Pl. 53:2), and only a narrow marginal part of the bone, close above the radius, was lost. After preparation this part was replaced with glue (Pl. 53:1).

The right humerus of this outstanding specimen is also well preserved but is still connected with the skull and is therefore only partly exposed (Pl. 46). It lies close below the lower jaw and with the caput humeri only a short distance behind the glenoid fossa of the shoulder girdle. The right radius and ulna, too, are present and have been completely freed from the matrix (Pl. 54).

Humerus. – It is evident that the humerus in *Ichthyostega* is a complicated structure, differing considerably from the humerus in the osteolepiform fish *Eusthenopteron* as well as from that in the Permian stegocephalian *Eryops* (Fig. 43B; Miner 1925; Jarvik 1964, Figs. 26–27) and other stegocephalians (see, e.g., Nilsson 1939, Figs. 6–8). It is composed of a proximal, rather thin *caput portion* and a *distal portion* including three thin laminae, one *dorsal* (the dorsal ridge), one *medial* (the entepicondyle), and one *lateral* including the ectepicondyle.

The outline and position of the oval articular area for the radius is shown on the left humerus (Pl. 53:2). On the right side of the specimen, the radius has been removed and the articular area on the right humerus is exposed (Pl. 46:1, 2, 4, 5). This area forms a fossa (*art.fo.R*) with an eminence (*cap*, Fig. 45F) which, although not spherical as the corresponding structure in man (*cap*, Fig. 43A), is most conveniently termed the *capitulum*. This articular fossa with the *capitulum* is situated on the ventral side of the posterior part of the distal lateral lamina of the humerus which I (Jarvik 1980a, p. 232) interpreted as the ectepicondyle. This interpretation has been criticized by Panchen (1985, p. 544), Smithson (1985, p. 365) and Clack (1987, p. 72), who like other students describing the humerus in late Palaeozoic stegocephalians (Holmes 1980; Godfrey 1989) have misinterpreted the ectepicondyle in *Ichthyostega*. It may be added here that the posterior part of the distal lateral lamina with the radial articular fossa is strengthened by a strong ridge, the *ectepicondylar buttress* (Fig. 44; Pl. 53:1, 3), which from the dorsal side of that part of the distal lateral lamina which represents the ectepicondyle, extends upwards on the lateral side of the dorsal ridge towards the dorsal end of the articular area for the ulna (*ect.buttr* Fig. 45A, B, D).

That the posterior part of the distal lateral lamina with the articular fossa for the radius is the true *ectepicondyle*, as I maintained in 1980, is evident if we compare with the ancestral fishes and post-Devonian eutetrapods. In *Eusthenopteron* as well as in *Eryops* (Fig. 43B; Jarvik 1964, Fig. 26; A, B; 1980a, Figs. 103, 104; Romer 1930, Fig. 135A) and other

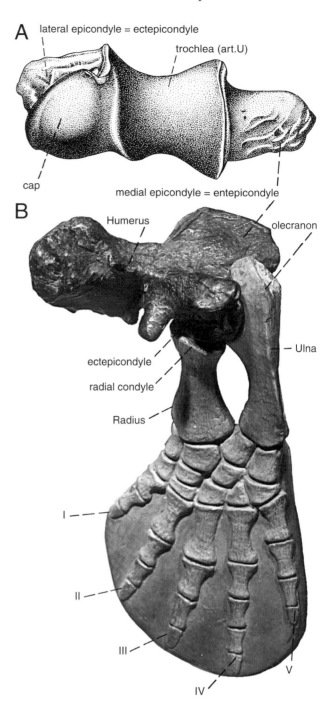

Fig. 43. □A. Distal end of left humerus of *Homo* seen from below. From Sobotta 1922, Fig. 118. □B. Forelimb of *Eryops*. Photograph of model in plaster in the Swedish Museum of Natural History *art.U*, articular area for ulna; *cap*, capitulum humeri; I, II, III, IV, V, digits *I–V*.

post-Devonian stegocephalians (Nilsson 1939), reptiles (Romer 1970, Fig. 135B, C; Guibé 1970, Fig. 54) and man (Fig. 43A), the ectepicondyle lies above the radius condyle (note that the ectepicondyle in man is called the 'lateral epicondyle'). Moreover, it is of interest to observe that the articular area for the ulna lies between the ectepicondyle and the entepicondyle.

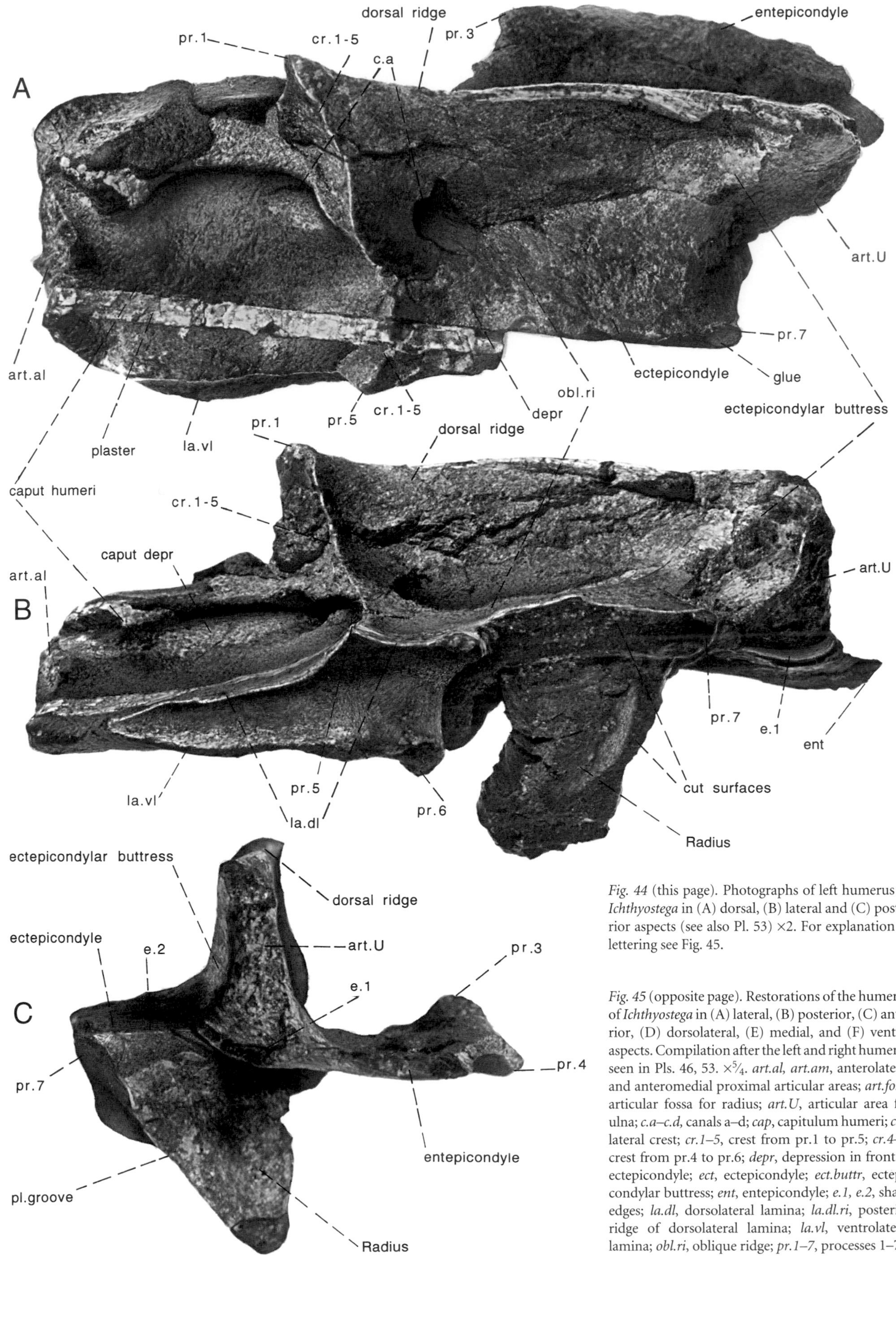

Fig. 44 (this page). Photographs of left humerus of *Ichthyostega* in (A) dorsal, (B) lateral and (C) posterior aspects (see also Pl. 53) ×2. For explanation of lettering see Fig. 45.

Fig. 45 (opposite page). Restorations of the humerus of *Ichthyostega* in (A) lateral, (B) posterior, (C) anterior, (D) dorsolateral, (E) medial, and (F) ventral aspects. Compilation after the left and right humerus seen in Pls. 46, 53. ×5/4. *art.al, art.am,* anterolateral and anteromedial proximal articular areas; *art.fo.R,* articular fossa for radius; *art.U,* articular area for ulna; *c.a–c.d,* canals a–d; *cap,* capitulum humeri; *cr.l,* lateral crest; *cr.1–5,* crest from pr.1 to pr.5; *cr.4–6,* crest from pr.4 to pr.6; *depr,* depression in front of ectepicondyle; *ect,* ectepicondyle; *ect.buttr,* ectepicondylar buttress; *ent,* entepicondyle; *e.1, e.2,* sharp edges; *la.dl,* dorsolateral lamina; *la.dl.ri,* posterior ridge of dorsolateral lamina; *la.vl,* ventrolateral lamina; *obl.ri,* oblique ridge; *pr.1–7,* processes 1–7.

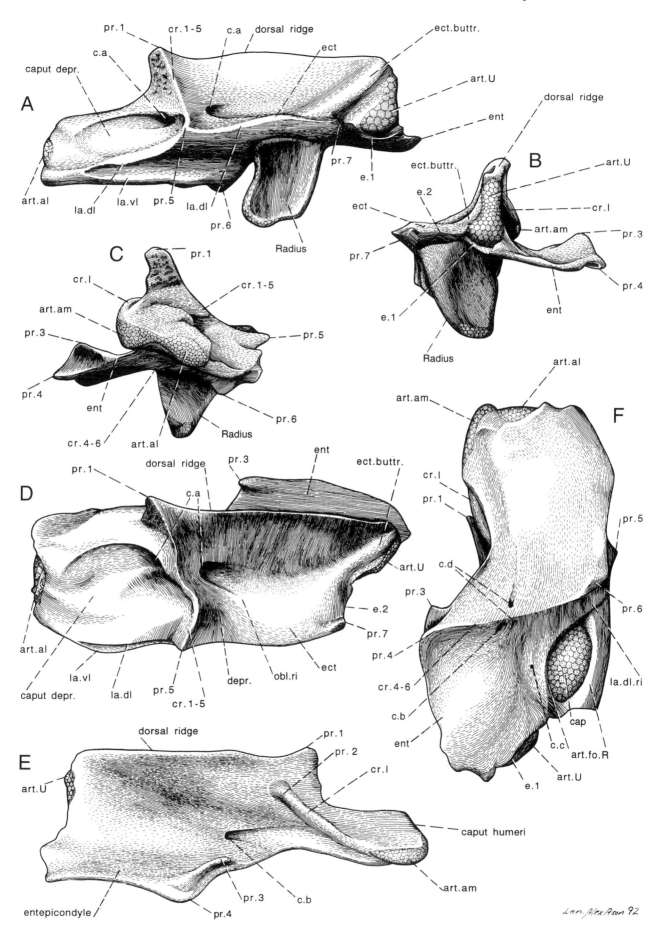

In *Ichthyostega*, the articular area for the ulna (*art.U*; Figs. 44, 45) is well shown both on the left (Pl. 53:1–3, 5) and the right (Pl. 46:1–3, 5) humerus. Its dorsal part is narrow, but it widens downwards, and its widened ventral end is bounded by a projecting thin edge (*e.1*), a condition which indicates that the area was capped by a thin layer of cartilage. The ulnar articular area in *Ichthyostega* is situated at the posterior end of the dorsal ridge, as it is in *Eusthenopteron*. However, as is well shown on the humerus of the outstanding specimen P.222 (Jarvik 1964, Figs. 25C, 26A), the dorsal ridge in *Eusthenopteron* is much lower than in *Ichthyostega*.

The articular area at the proximal (capitular) end of the humerus is shown both on the left (Pl. 53:2, 4, 5) and the right (Pl. 46:1, 4, 5) humerus. It includes (Fig. 45C; Pl. 53:4) a rounded anterolateral portion (*art.al*), which abuts against the anterior vertical part of the glenoid fossa, and a postero-medial somewhat triangular portion (*art.am*), which occupies the posterodorsal part of the glenoid fossa.

In addition to the proximal articular area, the articular areas for the radius and ulna, the dorsal ridge, the ectepi-condyle, and the entepicondyle (well preserved on the left side, Fig. 44A, C; Pl. 53), the humerus in *Ichthyostega* (Fig. 45) presents a number of processes (*pr.1–pr.7*), canals (*a–d*), and other structures unlabeled on my previous restorations (Jarvik 1980a, Fig. 166). The humerus shows many furrows, processes and foramina that have never been described previously. Their interpretation would require a detailed investigation of primitive living and fossil tetrapods, which is beyond the scope of this paper and for which I have no time. Therefore I will describe the structures without offering any anatomical interpretation, knowing that such an incomplete treatment, which is just the first step in recording the basic observations, will set the stage for much subsequent study.

Process 1, preserved only on the left humerus (Fig. 44A, B; Pl. 53: 1, 4, 5), is represented by the somewhat elevated anterodorsal corner of the dorsal ridge. Below this process the broadened anterior part of the dorsal ridge forms an anterolaterally facing area with a rough surface (Figs. 44A, B; 45A, C, D; Pl. 53: 1, 4, 5). Posterolaterally this area is bounded by a ridge, which on the lateral side of the bone is continued by a narrow arched crest (*cr. 1–5*) extending downwards and laterally to process 5 (Fig. 44A, B; Pl. 53:1, 4). This lateral arched crest, pierced by a canal (*ca*), forms the posterior border of the caput portion. In front of the crest, this portion presents a shallow depression (*caput depr.* Figs. 44B; 45A, D) bordered medially by a thick ridge. The concave, dorso-medially facing surface of that ridge extends ventrally to the posteromedial portion of the proximal articular area (*art.am*), and as is well shown on the right humerus (Pl. 46:1, 4, 5), this articular area is continued backwards by a crest (*cr.l*) on the lateral side of the dorsal ridge. Posteriorly this lateral crest, which ends with a free margin (process 2), is intimately connected with the dorsal ridge, and when the lateral crest is broken, as on the left humerus (Pl. 53:5), there is a perforation (*perf*) in the dorsal ridge which was mistaken

for the opening of a canal (canal *e*, Jarvik 1980a, Fig. 166B). Between the concave ventral surface of the lateral crest and the underlying marginal part of the bone there is a pocket opening backwards.

Farther back, the anterolateral corner of the entepicondyle presents another, but smaller, pocket bounded dorsally by the elevated process 3 (*pr.3*), and ventrally by process 4 (*pr.4*) of the entepicondyle (Pl. 53: 4, 5). The ventral side of process 4 presents the medial end of an *arched ridge* or crest (*cr. 4–6*; Fig. 45F; Pls. 46: 5; 53:2), which, passing in front of the radial articular fossa, extends to process 6. From that process (*pr.6*) a thin lamina (*la.vl*, Fig. 45A), formed by the ventrolateral marginal part of the bone, extends forwards along the caput humeri. A little behind the lateral part of the anterolateral portion of the proximal articular area (*art. al*), it meets a dorsolateral lamina (*la.dl*), which extends to process 5. (Figs. 44B, 45A). Between the dorsolateral and ventrolateral lami-nae there is, as is well shown on both humeri (Figs. 44B; 45A; Pl. 46:2), a depression or groove widening backwards to-wards the radial articular fossa.

Ectepicondyle and radial articular fossa. – Posterolateral to the wide posterior opening of canal (*a*) there is a low ridge (*obl.ri*; Figs. 44A; 45D) extending obliquely backwards from the posterior side of the lateral arched crest described above. Between this oblique ridge and the lateral arched crest (*cr.1–5*) there is, behind process 5, a depression (*depr.* Fig. 45D), bounded ventrally by the free lateral margin of the anterior part of the distal lateral lamina of the bone, which is a posterior continuation of the dorsolateral lamina (*la.dl*). This lamina extends backwards to process 7, ending at a small ridge ventral to that process (Fig. 44C; 45B; Pl. 53:3). Dorsal to the lateral side of the radius (where it is, as shown on Pl. 53:1, replaced by glue; see below), it forms the lateral border of the *ectepicondyle* (Figs. 44, 45; Pl. 53:1, 3). This structure, situated dorsal to the radial articular fossa, is an eminence on the posterior part of the distal lateral lamina continued posterodorsally by the ectepicondylar buttress described above. Anteroventrally the ectepicondyle is bounded by the low oblique ridge (*obl.ri*), in front of which the dorsal surface slopes steeply downwards into the depres-sion (*depr.*) mentioned above. Posteriorly the ectepicondyle ends with a sharp edge (*e.2*) situated between process 7 and the ventral part of the ulnar articular area (Fig. 45B; Pl.53:3).

The posterior part of the dorsolateral lamina (*la.dl*), from a little behind process 5 to process 7, was damaged by a saw cut (p. 59). In connection with the preparation of the speci-men, the part sawn away was replaced by glue (Pl. 53: 1) but later removed, exposing the cut surface, (Fig. 44B). Evidently it is this posterior part of the dorsolateral lamina bordering the ectepicondyle that forms the lateral wall of the *radial articular fossa*. It is then of interest to note that also the medial wall of that fossa is formed by the dorsolateral lamina. As is well shown on the right humerus (Pl. 46: 1, 2, 4, 5), this lamina is continued posteromedially by a strong ridge (*la.dl.ri*), which forms that wall. The posterior part of the fossa is obviously formed by the part of the bone that on the left humerus (Figs. 44C, 45B; Pl. 53:3) is exposed between the sharp edge (*e.2*) and the underlying proximal end of the radius and bordered laterally by the small ridge below process 7. The capitulum, well shown in the fossa, is a longitudinal eminence tapering towards both ends. It occupies a depression, the fovea, at the proximal end of the radius (Pl. 54:1, 4, 6).

According to my previous restorations (Jarvik 1980a, Fig. 166), the humerus of *Ichthyostega* is pierced by five canals (*a–e*). However, since one of these canals (*e*) is a perforation (*perf*, Pl. 53:5) on the left humerus that resulted from breakage of the lateral crest (*cr.I*) ending with process 2 (Pl. 46:4), only four canals can be recognized. The most prominent of these canals (*a*) pierces the lateral arched crest (*cr.1–5*, Figs. 44A, B; 45A, D). The remaining canals cannot be followed through the bone and have been recognized by their openings only.

Canal *b* (Fig. 45E, F) is a rather wide and short canal, well shown on the right humerus. The posterior opening of this canal is found close behind the arched crest 4–6, anteromedial to the medial wall of the radial articular fossa and close to the base of the entepicondyle (Pl. 46:3). It pierces that condyle and opens on its dorsal side close to the ventral part of the medial side of the dorsal ridge. This opening is well shown on the specimen but not on the photographs. On the left humerus the posterior opening of canal *b* is concealed by the radius (Pl. 53:3), whereas the anterior opening, continued forwards by a groove, is exposed (Fig. 45E; Pl. 53:5). Canal *c* is narrow and enters the bone through a small opening on the ventral side just medial to the radius (Pls. 46:1, 4; 53:2). The passage of this canal is unknown. Canal *d*, (Fig. 45F) finally, is a narrow canal which, as seen on the right humerus, enters the bone behind the arched crest 4–6, a little anterolateral to the posterior opening of canal *b* (Pl. 46:1, 4, 5), and is continued forwards by a groove on the ventral side of the bone.

Radius. – The left radius in A.115, (Fig. 44; Pl. 53), attached to the humerus, is preserved in its natural position. A narrow posterolateral strip of the bone has been sawn away (the cut surface is seen on Fig. 44B and Pl. 53:1), but otherwise it is complete and well preserved. The right radius of the same specimen (Pl. 54:1–7), also well preserved, was detached at the preparation but may be put back in the radial articular fossa of the right humerus (Pl. 46).

The radius in *Ichthyostega* (Fig. 46A–C) is a relatively short and stout, three-sided bone suggestive of the radius in *Eryops* (Fig. 43). It differs from the more slender types found in *Archeria*, described in detail by Romer (1957), and from Carboniferous stegocephalians such as *Proterogyrinus* (Holmes 1980), *Crassigyrinus* (Panchen 1985), *Pholiderpeton* (Clack 1987) and *Greererpeton* (Godfrey 1989). Of the three main sides of the bone, all with smooth concave sur-

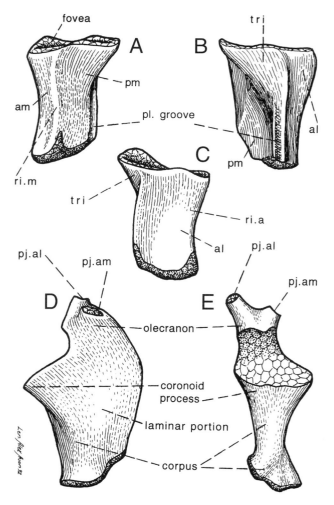

Fig. 46. Drawings of right radius and ulna of *Ichthyostega* after the examples shown in Pl. 54. □A–C. Radius in medial, posterior and lateral aspects. □D, E. Ulna in medial and anterior aspects. *al, am,* anterolateral and anteromedial surfaces; *pj.al, pj.am,* anterolateral and anteromedial projections; *pl.groove,* posterolateral groove; *pm,* posteromedial surface; *ri.a, ri.m,* anterior and medial ridges; *tri,* triangular area.

faces, the anterolateral one meets the anteromedial side (*am*) at the anterior rounded margin (*ri.a*), which is concave as seen in lateral and medial views (Pls. 53:1; 54:1, 3, 6). The anteromedial and posteromedial sides meet in a rounded medial ridge (*ri.m*). A remarkable feature is the presence of a groove (*pl.groove*) at the posterolateral end of the bone. This groove, which in the distal half of the bone is situated between the anterolateral and posteromedial sides, merges proximally into a triangular area (*tri*) with a concave surface (Pl. 54:5). At the proximal end of the bone there is a depression or fovea to receive the capitulum of the radial articular fossa.

Ulna. – This element in *Ichthyostega* (Fig. 46D, E) is represented only by the right ulna of A.115 (Pl. 54:8–12). It has been freed from the hard embedding sandstone, and the well-preserved bone can be studied in various aspects. It is

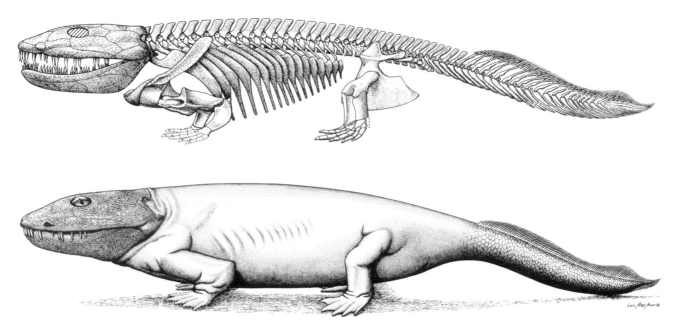

Fig. 47. Restorations of *Ichthyostega* in lateral aspect. For other restorations, see Bjerring 1988 and Klembara 1991.

already a typical tetrapod ulna and, as in *Eryops* (Fig. 43), distinctly longer than the radius. It includes a thickened corpus with a concave anterior margin and a rather large posterodorsal laminar portion, with concave lateral and medial surfaces. The ulna in *Ichthyostega* is thus much broader in lateral or medial views than that of *Eryops* (Fig. 43), *Archeria* (Romer 1957) and other late Palaeozoic stegocephalians. At the proximal end it presents a semilunar area for the articulation with the humerus, suggestive of the corresponding area in man (see, e.g., Gray 1973, p. 209, Fig. 4-139), being bounded by two processes, one ventral, the *coronoid process,* and one dorsal, the *olecranon.* A characteristic feature of the ulna in *Ichthyostega* is that the olecranon is provided with two projections, one anteromedial (*pj.am*) and one anterolateral (*pj.al*). At the distal end a small bone, probably a carpal, is retained, but no other carpals have been found and the digits are unknown.

Pelvic girdle and hindlimb. – Each *pelvic girdle* in *Ichthyostega* (Figs. 47–49; Pls. 55–62) is represented by a well-ossified, purely endoskeletal bone (cf. Panchen & Smithson 1990), which, judging from conditions in post-Devonian stegocephalians such as *Archeria* and *Eryops* (see, e.g., Romer 1970, Fig. 126), includes iliac, pubic and ischiadic portions (cf. *Eusthenopteron,* Jarvik 1980b, Fig. 82). In A. 93 (Pl. 55) and A. 250 (Pl. 56), the right pelvic bone has been removed from the matrix and can be studied in various aspects (Pl. 59). No traces of sutures between the three components are discernible in these or most other specimens. It is then of interest that the pubis in A. 163, (Fig. 48C; Pl. 62:3) and in two other specimens (Pls. 57:1; 63:1) is separated from the ischiadic portion by a distinct but incomplete suture (see below).

The iliac portion obviously forms the posterodorsal part of the acetabulum. Dorsal to that fossa is a rounded neck formed by thick bone and also as in *Archeria* (Romer 1957, p. 113) a shovel-shaped blade with a well-developed posterior *iliac process*. A feature characteristic of *Ichthyostega* is that the neck is pierced by a narrow iliac canal (*c.il*), the openings of which are shown in several specimens (Pls. 55–58; 59:3, 4; 60:1; 61:1–3; 62:3). The outer side of the blade, which is separated from the neck by a ridge, is concave, and the blade thins towards its anterodorsal margin. On the inner side of that marginal part of the blade is a depression, which has a rough surface for the articulation with a sacral rib. At the dorsal end of that depression there are two small processes, one anterior and one posterior (*pr.a, pr.p,* Figs. 48B; 50A; Pls. 55, 56, 58:1, 59:2–4). Most likely these processes, which lack periosteal lining at the top, are for the attachment of ligaments connecting the pelvis with the sacrum.

The prominent posterior iliac process, which possibly (Jarvik 1952, p. 14) is formed by the incorporation of a postsacral rib, is represented in A. 93 (Pls. 55, 59) only by its proximal part; it is more or less well shown in Pls. 56–58, 61 and 62:3. As in *Archeria,* it may bear a longitudinal striation.

The main part of the pelvic bone in *Ichthyostega* is formed by the extensive puboischiadic plate. On the inner side of that plate, well shown in particular in A. 93 (Fig. 49A; Pls. 55:2, 59:3), there is a broad rounded ridge extending downward to the highest point of the well-developed symphysial area. This ridge obviously corresponds to 'the prominent ridge', which also in *Archeria* (Romer 1957, Fig. 3, p. 114) 'divides the inner surface of the bone into two distinct areas, an anteromedial triangle leading down onto the inner surface of the pubis, and a broader surface facing inward and slightly posteriorly, continued by the ischium'. Other similarities between *Ich-*

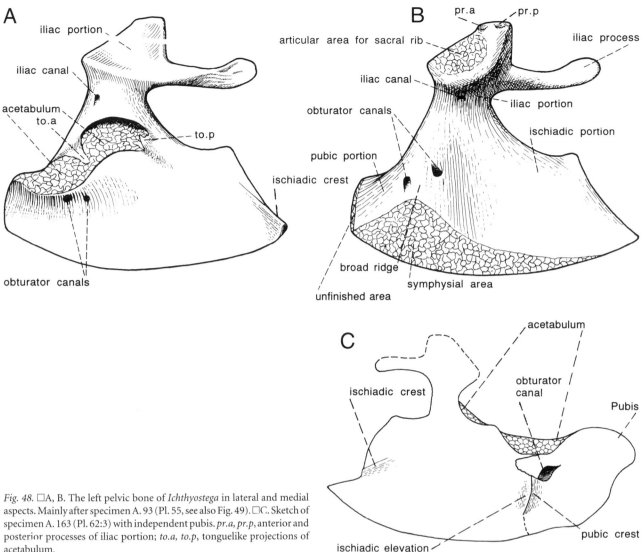

Fig. 48. □A, B. The left pelvic bone of *Ichthyostega* in lateral and medial aspects. Mainly after specimen A. 93 (Pl. 55, see also Fig. 49). □C. Sketch of specimen A. 163 (Pl. 62:3) with independent pubis. *pr.a, pr.p,* anterior and posterior processes of iliac portion; *to.a, to.p,* tonguelike projections of acetabulum.

thyostega and Archeria are that the pubic portion is short and the anterior vertical margin lacks periosteal lining (Fig. 48B), indicating that there was 'an anterior cartilaginous continuation of the pubis' (Romer 1957, p. 115). This may be of some interest, since the portion of the pelvic bone in *Eusthenopteron* (Jarvik 1964, Fig. 24; 1980b, Fig. 82) that – according to Gregory & Raven (1941), Jarvik (1964), and others – represents the pubis, is a long rod, longer than the portion that in my opinion represents the ischium. Note that the dorsal margin of this portion in *Eusthenopteron* is concave and suggestive of the dorsal margin of the ischiadic portion in *Ichthyostega* (Fig. 48A, B).

As described by Romer (1957, p. 115), the inner surface of the pubis in *Archeria* is interrupted by the inner opening of the obturator canal ('foramen'), and he adds that 'a rounded, grooved channel in the bone leads down to it from above'. This is of interest, since in specimen A. 93 (Figs. 48B, 49A; Pl. 55:2), in which the pubic portion is pierced by two obturator canals, a similar 'grooved channel' from the inner

opening of the anterior canal, exactly as in *Archeria*, leads down to it from above. A similar 'grooved channel' is associated with the inner opening of the posterior obturator canal, but this 'channel' leads upwards to the opening from below. In A. 250, too, there are two obturator canals (Pl.56). This is obviously the case also in A. 97, in which openings of two canals in the same position as in A. 93 are shown on the inner side of the bone (Pl.58:1). In A. 155 (Fig. 49C; Pl. 60), two obturator canals enter the bone from the inner side but unite and have a common large opening on the outer side.

The obturator openings on the outer side (Figs. 48A, C; 49B–E) are situated immediately below the elongated acetabulum, in the posterior part of the pubic portion. The deep acetabular fossa extending forwards almost to the anterior margin of that portion lacks periosteal lining except in two tongue-like areas, one anterior (*to.a*) and one posterior (*to.p*), (Fig. 48A; Pls. 55:1; 56:1; 57; 60:1). Similar areas, both situated on the ilium, are found also in *Archeria* (see Jarvik 1980a, Fig. 161A).

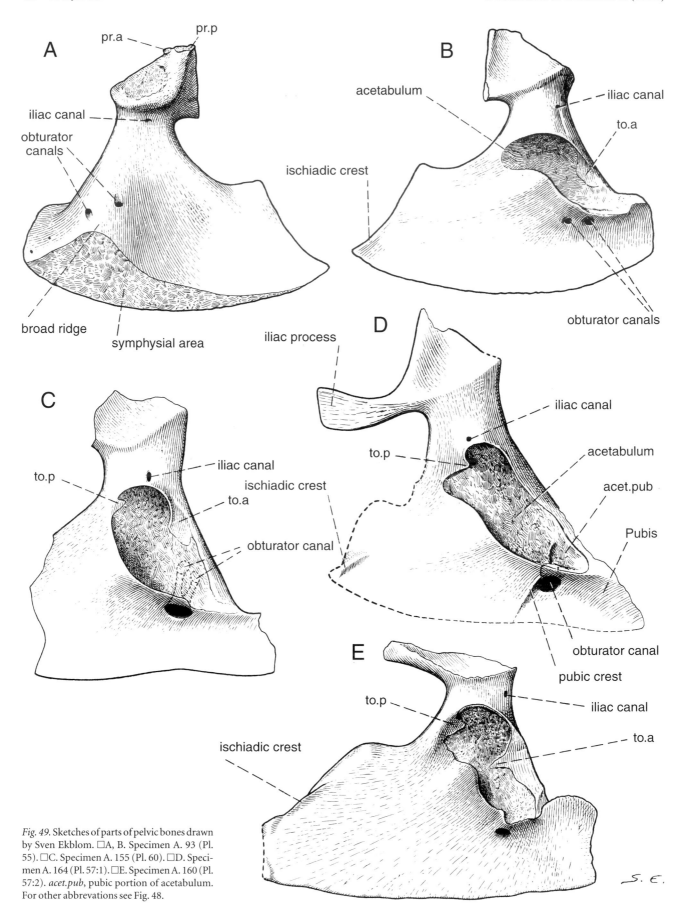

Fig. 49. Sketches of parts of pelvic bones drawn by Sven Ekblom. □A, B. Specimen A. 93 (Pl. 55). □C. Specimen A. 155 (Pl. 60). □D. Specimen A. 164 (Pl. 57:1). □E. Specimen A. 160 (Pl. 57:2). *acet.pub*, pubic portion of acetabulum. For other abbrevations see Fig. 48.

In addition to the acetabulum and the openings of the iliac and obturator canals, the outer side of the pelvic bone in *Ichthyostega* presents two structures not previously recorded. In A. 163 (Fig. 48C; Pl. 62:3) the dorsal part of the pubic portion, dorsal to the obturator canal, projects backwards outside the ventral border of the acetabulum formed by the ischiadic portion. This posterior pubic projection, which has a concave outer surface, ends posteroventrally with a free margin continued forwards by a distinct arched crest extending downwards almost to the ventral margin of the bone. Posteriorly this *pubic crest* meets an elevation on the anterior part of the outer side of the ischiadic portion. Between this elevation and the crest there seems to be a suture continuing upwards to the acetabulum. In A. 164, (Pl. 57:1), in which a similar pubic crest is present (Fig. 49D), it is continued by a low ridge in the bottom of the acetabulum, between the anterior pubic (*acet.pub*) and the posterior ischiadic portions of that fossa. These conditions indicate that the pubis is independent or, at any rate, that we are witnessing an incipient subdivision of the pelvic bone into three independent bones as in later tetrapods (see also Pl.63:1, 2).

The other structure on the outer side of the pelvic bone is a prominent crest found on the posterior part of the ischiadic portion. The dorsal part of this *ischiadic crest* is shown on the two specimens (A. 163, and A. 164, Pls. 57:1, 62:3) mentioned above but is more completely shown in A. 93 (Fig. 49B; Pl.55) and, in particular, in A. 160 (Fig. 49E; Pl. 57:2). In that specimen it forms a sharp ridge extending downward to the posteroventral corner of the ischiadic portion, where it seems to end with a process lacking periosteal lining at the top (see Fig. 48A).

Remarks on polydactyly and metapterygial stem. – Pentadactyly has long been considered to be a characteristic feature of tetrapods. The discovery (Lebedev 1984, 1990; Lebedev & Clack 1993) of six digits in the forelimb of a Devonian tetrapod, *Tulerpeton*, therefore attracted some interest. However, in the Indian family Koli Patel many members have six fingers with well-developed musculature on each hand (Fig. 50). There is no information about the toes, but Professor J. Slípka, Plzeň, has sent me photographs (Fig. 50) of limbs in aborted human fetuses showing six digits both in

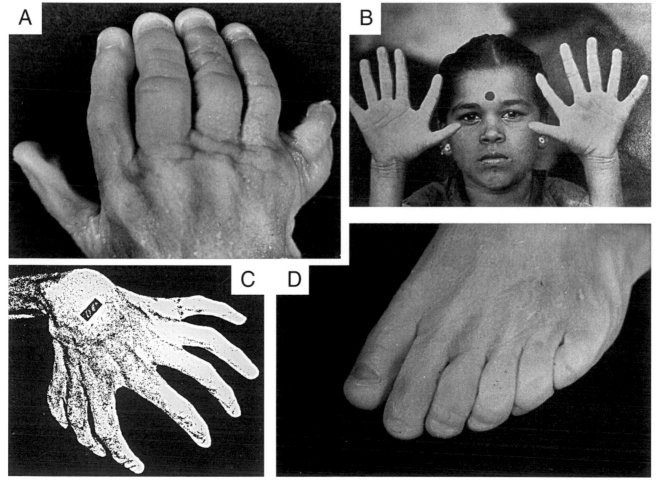

Fig. 50. Some examples of polydactyly in man. □A, D. Photographs of male human foetuses about 24 weeks with (A) six fingers and (D) six toes, presented by Professor J. Slípka, Plzeň. □B. Girl of the Indian family Koli Patel with six well-developed fingers on each hand. From the Norwegian magazine Hemmet 18/91. □C. Aberrant human hand with seven fingers. The forearm consists of the normal left ulna and of a right one in the place of the radius. From Dwight 1892, p. 477, Pl. 43:2.

the hand and the foot. He also added X-ray photographs showing skeletal elements in the digits. Also many other cases of polydactyly in man e.g., of the hand in a southern Chinese and a seven-toed foot have been recorded (for references see Gray 1973, p. 284; see also Bergglas 1925). Since polydactyly occurs in man it may be regarded as strange that it lasted until 1984 before this phenomenon was recorded in the paleontological literature. More recently, Coates & Clack (1990), on the basis of new material of Devonian tetrapods from East Greenland, have described polydactyly and claim that the forelimb of *Acanthostega* has eight digits and the hindlimb of *Ichthyostega* seven (see also Coates 1994, Tabin 1992). In their interpretations they are strongly influenced by studies mainly of one of the most specialized tetrapod limbs, the wing of birds (Hinchliffe & Johnson 1980), and they accept the view that in the tetrapod limb there is a digital arch homologous with the metapterygial axis (Shubin & Alberch 1986, p. 370).

According to current views (Jarvik 1965b, pp. 144–145, Fig. 2; see also above), the elusive metapterygial stem or axis is a row of metameric skeletal elements (basals) originally situated in the body wall, but which later by successive deepening of a posterior embayment (fissura metapterygii) has become freed from that wall and form the stem of the paired fins and the tetrapod limb. However, on the basis of Müller's (1909, 1911) detailed studies of the pectoral fin in sharks and its ontogenetic development in *Acanthias* (*Squalus*), Sewertzoff's studies (1926a, b, 1934) of the development of the paired fins in *Acipenser*, paleontological data provided by Stensiö (1959), and other sources (Goodrich 1930, with references), it could be demonstrated (Jarvik 1964, 1965a, 1965b, 1980b) that the notion of a metapterygial stem, or axis, rests on a misinterpretation of the pectoral fin in sharks; I concluded (Jarvik 1965b p. 166) that 'there is no metapterygial stem either in fishes or in tetrapods'.

The muscles and other soft parts are most important components of the paired fins and the tetrapod limbs. As is well known in fishes (Müller 1911, Goodrich 1930, and others), the muscles of the paired fins are derived from the ventral ends of the growing metameric myomeres where *muscle buds* are formed. Some of these muscle buds may disappear (abortive buds), but those remaining form *radial muscles*, which grow in the distal direction towards the margin of the fin fold. Two dorsal and two ventral radial muscles arise in each metamere. To form a support for the radial muscles, a skeletal rod, generally referred to as *radial* or *ray*, arises between each dorsal and ventral radial muscle. Together with their muscles, the rays grow towards the margin of the fin, but a long overlooked fact is the presence of a secondary muscle bud ('a small process-like outgrowth'; Jarvik 1965b; p. 160, *rmp*, Fig. 9) at the dorsal end of each primary muscle bud, where the pterygial nerves enter the latter bud. The secondary portions of the radial muscles arising from the secondary buds grow together with secondary portions of radials (rays) in a proximal direction towards

the shoulder joint (Fig. 12). By a well-established morphogenetic process, the base of the fin becomes shortened from behind, which results in a crowding together and fusion of the proximal portions of the secondary radial portions (Goodrich 1930, p. 132, Fig. 156). This is what students of chicken limbs (see e.g., Hinchliffe 1977, p. 305) refer to as 'a bifurcating system'. It is also evident that the basale metapterygii in sharks, which is an important part of the 'metapterygial stem', is formed by fusion of adjoining proximal portions of radials (Fig. 12) and is a secondary element arisen in the free fin, not a metameric element freed from its original position in the body wall.

In tetrapods, the limb muscles are also formed by myogenic material emerging from the ventral ends of the growing myotomes. Primary muscle buds are known in many forms and are well shown in the reptile *Ascalabotes* studied by Sewertzoff (1904, 1908, see Jarvik 1980b, p. 135, Fig. 80; note also the implantation experiments in urodele larvae carried out by Balinsky 1933, 1935, 1937; see also Jarvik 1965b, pp. 146–148; 1980b, pp. 115–116). As regards the limb muscles, there are thus no fundamental differences between fishes and tetrapods, and no doubt the skeletal elements in the tetrapod limbs arise in association with radial muscles, as in fishes. In tetrapods no radial muscles are discernible, but there are dorsolateral and dorsomedial muscle primordia (Romer 1944; Jarvik 1965a, p. 132; 1980b, pp. 135–136). In fishes the number of radials varies, but in, e.g., *Acanthias* (*Squalus*, Jarvik 1965b, Fig. 6) there are 24 pairs of radial muscles and 24 radial rays in the pectoral fin. How many radials that enter into the formation of the eutetrapod limb is difficult to say. There may be as many as recently suggested by Bjerring (1985, 1988), but if we keep to the distal ends, the limbs in man are seven-rayed as are the pectoral and pelvic fin in the ancestral fishes (*Eusthenopteron*, Jarvik 1965a, Fig. 10; 1980b, Figs. 81, 83).

Before proceeding to my material of the hindlimb in *Ichthyostega,* it may be mentioned that Shubin & Alberch (1986) contrasted two modes of tetrapod limb development, one seen in larval salamanders and the other in all remaining tetrapods. This is only another example of important differences we always find when comparing urodeles with eutetrapods and supports my contention that the Amphibia are diphyletic (Jarvik 1980a, 1986).

Hindlimb. – Remains of the hindlimb in *Ichthyostega* are shown in several specimens (Pls. 63–68; also Jarvik 1952, Fig. 5), but A. 109 (Pls. 65:1, 2; 66) and A. 166 (Pls. 63, 64) are of special interest in that they both exhibit five digits (I–V). Photographs of the hindlimb in A. 109, published several times since 1952, show the ventral side of the left hindlimb with femur, tibia, intermedium, fibulare, and the pentadactyl foot. The femur and tibia have been removed and are presented in various aspects (Pl. 66). In A.166 the left foot is separated from the less well-preserved right foot (not figured) by the pelvic bones (Pl. 62:1). The counterpart of the

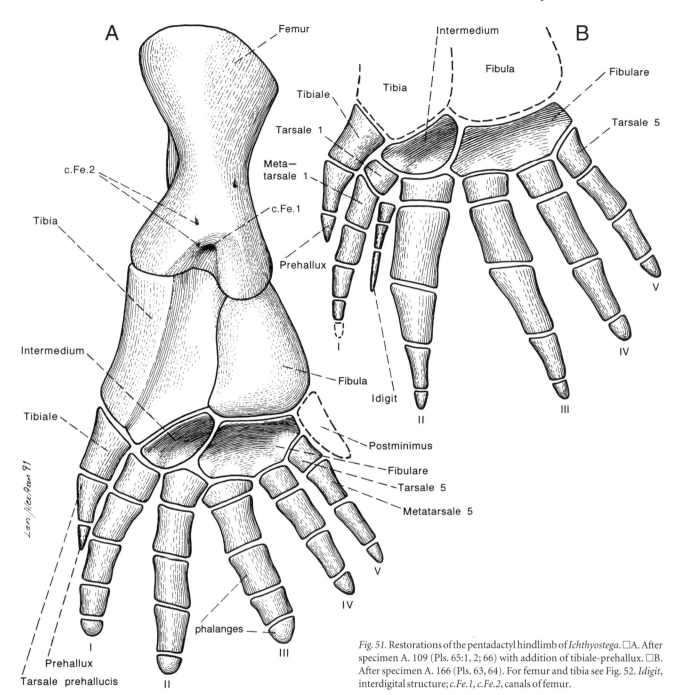

Fig. 51. Restorations of the pentadactyl hindlimb of *Ichthyostega*. □A. After specimen A. 109 (Pls. 65:1, 2; 66) with addition of tibiale-prehallux. □B. After specimen A. 166 (Pls. 63, 64). For femur and tibia see Fig. 52. *Idigit*, interdigital structure; *c.Fe.1, c.Fe.2*, canals of femur.

left foot is well preserved, and it has been possible to make thin latex casts showing both sides of the foot (Fig. 52; Pls. 63;1, 64:1, 2, 4, 5). The fibula is not preserved, but there are remains of the femur and the tibia (Pls. 63; 64:1, 2, 5). The latter is followed by a well-preserved tibial ray including tibiale, tarsale prehallucis and prehallux (Pls. 63; 64:2–3). Intermedium, fibulare, tarsals, metatarsals, and digits I–V are present as well, but a remarkable fact that has puzzled me for a long time is the presence of a narrow two-jointed rod between digits I and II (Fig. 51B; Pls. 63; 64:1–3). However, recently Hurle & Ganan (1986, p. 242) recorded 'cell death,

often accompanied by the differentiation into cartilage of the interdigital mesoderm' and stated (p. 236): 'In five cases, the ectopic cartilages appeared as two pieces separated by a developing joint'. In view of this evidence, the two-jointed rod between digits I and II is to be regarded as an *interdigital* structure (*Idigit*), probably developed to strengthen the web in this mainly aquatic animal. A similar interdigital structure in a corresponding position is present also in the specimen of *Ichthyostega* currently available for studies in Cambridge (Coates & Clack 1990, Fig. 1 d–f). The view that the hindlimb in *Ichthyostega* has seven digits is therefore to be rejected, and

Fig. 52. Photograph of cast of specimen A. 166 showing parts of femur, tibia and foot (cf. Pls. 63:1, 2; 64:1, 2, 4, 5). *art.Tib*, area of tibia articulating with tibiale; *intertr.fossa*, intertrochanteric fossa; *int.tr*, internal trochanter; *prox.art*, proximal articular area; *tr.4*, fourth trochanter.

comparison with the ancestral fishes shows that pentadactyly is primitive in the eutetrapod limbs (cf. Tabin 1992; Gould 1991, 1993, pp. 63–78). As is well established in *Eusthenopteron*, the endoskeleton in both the pectoral and pelvic fins is seven-rayed (Fig. 12; Jarvik 1980b, Figs. 81A, 83A). If we connect the distal ends of the first and seventh rays with a straight line, we will find that the distal ends of rays 2–6 reach beyond this line. When, at the transition from fish to tetrapod, the animals began to use the paired fins for walking, it is evident that it was the distal ends of rays 2–6 that first touched the ground. It is therefore natural that these five rays were utilized to form digits I–V in the pentadactyl hand and foot, whereas the first and seventh rays formed the prepollex and pisiforme in the hand and the prehallux and postminimus in the foot. It is also easy to understand that the main load of the body was carried by the side of the prepollex in the hand and the prehallux in the foot, and that it was this side of the web that needed support. Possibly this explains why the interdigital structure both in A.166 and in the Cambridge specimen of *Ichthyostega* arose between digits I and II. In this connection it may be mentioned that the distal phalanges of digits I–V are hoof-like, in contrast to those of the prehallux and the interdigital structure, which are pointed. What this difference means is difficult to say, but the possibility cannot be excluded that the real digits in the foot of *Ichthyostega* were equipped with claws.

An important fact, well shown in the *Ichthyostega* specimen under study in Cambridge, is that the intermedium is bent at the transition between the vertical lower leg and the horizontal foot, which was planted on the ground when the animal was walking. Although less well shown, the intermedium is bent also in A. 109 and A. 166, but it is of interest that the fibulare, too, is distinctly bent in a similar way in the latter specimen (Pl. 64:5).

Femur. – Disregarding the proximal articular area, the femur (Fig. 53A, B) is fairly well preserved in A. 109 (Pls. 65:1; 66) and is found isolated in some other specimens (Fig. 52; Pls. 67, 68). It is a stout bone, similar to that in *Archeria* (Romer 1957, Fig. 8) and other early tetrapods (Romer 1956b, Fig. 170; 1970, Fig. 142), with a well-developed proximal articular area (*prox.art*) well shown in A. 77 (Pl. 67:2, 5) and A. 166 (Fig. 52). The smooth, gently convex dorsal surface of the bone presents the openings of two femoral canals (*c.Fe.*1; *c.Fe.*2; Pl.66:1), both leading into the intercondylar groove or fossa (*interc.fossa*, Fig. 53) at the distal end of the bone. This fossa is bounded by two prominent ridges (or condyles), one posterior, the fibular ridge (*ri.Fi*), ending with the articular area for the fibula (*art.Fi*) and one anterior, the tibial ridge (*ri.Ti*), with the rounded main portion of the articular area for the tibia. The latter area (*art.Ti*) is continued by a narrow portion at the distal margin of the *intercondylar fossa* (Pl.67:1, 3, 6). The ventral side of the bone shows proximally the *intertrochanteric fossa* (*intertr.fossa*) as in *Archeria*, with small foramina in its bottom (Pl.68:2, 3). Anteriorly this fossa, ventral to the proximal articular area, is, as seen in A. 166 (Fig. 52), bounded by a ridge with small tubercles, which probably (cf. Romer 1970, Fig. 142) is the internal trochanter (*int.tr*); ventral to that lies a rectangular rough area, which may be the fourth trochanter (*tr.4*). Distal to that area follows a distinct oblique ridge, which obviously is what Romer (1957, p. 132) calls the 'adductor crest' (*add.cr*) and which bounds the popliteal space (*pop*, Fig. 53).

Tibia. – In *Ichthyostega*, the tibia (Figs. 52; 53C, D; Pls. 63:1, 2; 65; 66; 68:5, 7) is a broad, robust structure differing considerably from that in, e.g., *Archeria* (Romer 1957, Fig. 9), *Eoherpeton* (Smithson 1985, Fig. 30A–E), *Greererpeton* (Godfrey 1989, Fig. 25g–s) and other eutetrapods (Romer

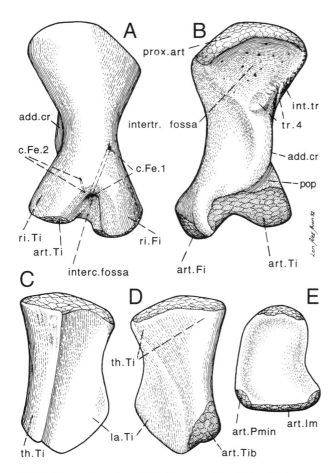

Fig. 53. Restorations of proximal elements of hindlimb of *Ichthyostega*. □A, B. Left femur in dorsal and ventral aspects. □C, D. Left tibia in dorsal and ventral aspects. □E. Left fibula in ventral aspect. *add.cr,* adductor crest; *art.Fi, Im, Pmin, Ti, Tib,* articular areas for fibula, intermedium, postminimus, tibia and tibiale; *c.Fe.1, c.Fe.2,* canals 1 and 2 of femur; *interc, intertr,* intercondylar and intertrochanteric fossae; *int.tr,* internal trochanter; *la.Ti,* laminar portion of tibia; *pop,* popliteal space; *prox.art,* proximal articular area of femur; *ri.Fi, ri.Ti,* fibular and tibial ridges; *th.Ti,* thickened portion of tibia; *tr.4,* fourth trochanter.

1956b, Fig. 177). It is composed of a thickened anterior portion (*th.Ti*), articulating with the tibiale (*art. Tib*), and a posterior laminar portion (*la.Ti*) articulating with the intermedium (*art.Im*, Pl. 64:4, 5). A prominent feature on the dorsal or extensor side is a sharp edge of the thickened portion extending from the proximal to the distal end of the bone. Proximally the thickened portion bears a rugose area. The dorsal side of the laminar portion presents a concavity continued upwards by the femoral intercondylar fossa (Pl.66:1). Proximally, on the broadened portion, the ventral (flexor) side of the bone close to the proximal articular area (Pl. 66:2) shows a depression (*depr.Ti*). At the transition to the laminar portion, this is bounded by a slightly arched ridge extending downwards to the distal end of the bone, close to the articular area for the tibiale.

Fibula. – This element (Fig. 53E) is preserved only in A. 109 (Pl. 65:1, 2), in which the ventral (flexor) side is exposed (for

the convex dorsal side, see Coates & Clack 1990, Fig. 1e). In contrast to conditions in other early tetrapods, it is a broad rectangular plate. The concave ventral (flexor) side is bounded posteriorly by a ridge terminating distally with a rounded articular area (*art.Pmin*) for the postminimus.

Final remarks

The main purpose of this paper has been to describe and photographically record the large material of *Ichthyostega* collected since 1931 in central East Greenland by members of the Lauge Koch expeditions. In May 1950, when I was free to start the description of this material, it was known that the ichthyostegids belong to a sidebranch of the Tetrapoda, in some respects primitive, in others specialized.

However, before turning to the ichthyostegids, I found it most urgent to go on with my studies of the two groups of Devonian fishes, the Osteolepiformes and the Porolepiformes, among which, as I have claimed since 1942, the former include the ancestors of the majority of the Tetrapoda. In the years 1950–1980 I therefore devoted almost all my time to further explorations of these two groups. For that purpose I had at my disposal an excellent fossil material as well as serial sections made according to Sollas's grinding method of the skull in one osteolepiform (*Eusthenopteron*) and (from 1959) one porolepiform (*Glyptolepis*). Following strictly the rules of comparative anatomy, the application of this unsurpassed method rendered it possible to make out in detail the cranial anatomy including nerves and other soft parts in osteolepiforms and in important respects also in porolepiforms. By this a solid basis was created of importance for comparisons with *Ichthyostega* and other stegocephalians, and, as must be strongly emphasized, indispensable for discussions of relationships.

Unfortunately, no skull of *Ichthyostega* suitable for investigation by serial grinding has been found, and therefore no detailed description of the cranial anatomy can be given. However, studies of skulls prepared mechanically, and the well-preserved postcranial skeleton, have shown that the ichthyostegids are related to the osteolepiforms and belong to the Osteolepipoda. The fact that an extrascapular series of osteolepiform type is retained in man refutes the view that the extrascapulars are lost in tetrapods, and the widespread Westoll–Romer terminology of the dermal bones of the skull roof, resting on a false basis, is rejected. In *Ichthyostega,* the extrascapulars are retained but have fused with the parietals and supratemporals. Other specializations concern the strange processes and laminae of the exoskeletal skull roof penetrating or bordering the otoccipital. Moreover, the palatoquadrate is provided with a parotic crest, which occupies a deep parotic groove in the lateral side of the otoccipital and is connected with the dorsal end of one of the exoskeletal processes. These facts, the solid skull table and the structure of the palate show that the ichthyostegid skull is akinetic, as is that of *Eusthenopteron*. The ventral position of the external nostril together with inturning of the anterior end of the maxilla are specializations unparalleled in later tetrapods. Attachment areas for a paired basicranial muscle spanning the fissura preoticalis are identified. In this respect the ichthyostegids are primitive. Another primitive feature is the long dentary, which extends backwards to the jaw joint. Yet another primitive character is the presence of an independent arcual plate, ventral to which the notochord, tapering in size, extended forwards to a small notochordal pit close behind the fossa hypophysialis. In contrast, in the anterior extent of the maxilla and in the shift in the biting function from the inner to the outer dental arcades, they have reached the tetrapod level.

In connection with the description of the vertebral column in *Ichthyostega,* its cranial portion is analysed and the misleading term 'resegmentation' is discussed. Occipital condyles and canals for occipital arteries, situated as in *Eusthenopteron* and Recent anurans, are present.

The ribs are unique in possessing a broad laminar portion, covering the succeeding rib, and canals for cutaneous blood vessels piercing its dorsal part. This fact, together with the weak squamation, indicates that *Ichthyostega* was adapted to skin breathing as a supplement to lung breathing. The fish-like tail is a primitive character, but it may be noticed that the ventral side of the tail is modified in a way that suggests that it was dragged on the ground when the animal was walking. In its appendicular skeleton *Ichthyostega* has reached the tetrapod level, characterized by true digits, progressive development of the endoskeletal girdles, and partial reduction of the exoskeletal shoulder girdle. The well-preserved endo- and exoskeletal shoulder girdles, humerus, radius and ulna, are described. The misinterpretation of the ectepicondyle is corrected, and it is shown that this structure with the radial articular fossa on its ventral side is strengthened by a massive ectepicondylar buttress. Carpals and digits in the forelimb are unknown. The pelvic girdle is a strong endoskeletal bone meeting its fellow in a median symphysis and joined to the vertebral column by ligaments and one or more sacral ribs. The pubis is sometimes independent. The hindlimb, well preserved in two specimens, is pentadactyl. However, in addition to the five digits, the pes includes a prehallux and a postminimus and is thus seven-rayed, as is the endoskeleton of both the pectoral and pelvic fins in the osteolepiform *Eusthenopteron*. Moreover, between digits I and II in one of the pedes, there is a narrow jointed rod, interpreted as an interdigital structure arisen to support the web.

Acknowledgements. – This paper is dedicated to the memory of the late Professors Erik Stensiö and Gunnar Säve-Söderbergh. They recommended me to be a member of the Danish East Greenland expeditions and introduced me to the collection and study of Devonian vertebrates. I also wish to remember the late Dr. Lauge Koch for his generosity and for the excellent way in which he facilitated the collecting work during my eight summers in East Greenland. My thanks are due to the late Professor Tor Ørvig and Professors Valdar Jaanusson and Jan Bergström, after my retirement direc-

tors of the Department of Palaeozoology, Swedish Museum of Natural History, where this paper has been prepared. Thanks are also due to members of the staff of this institute, particularly to Miss Eva Norrman, who after Miss Agda Brasch has continued the preparation of the fossils and made numerous latex casts, Mr Uno Samuelson who has taken most of the photographs in this paper together with the many photographs that have served as a basis for drawings, Mr. Bertil Blücher for retouching and mounting the photographs in the plates, Mr. Lennart Alex Andersson for his interest and insight into my work when drawing the many new restorations and for labelling the text-figures and plates, Mrs. Sif Samuelson for typing the manuscript, and Miss Ingela Chef-Johansson and Mrs. Ritva Woode for assisting with the final preparation of the manuscript. I am most grateful to Professor Jaroslav Slípka, Plzeň, for sending me photographs of limbs of aborted human foetuses. The editor of the series, Dr. Stefan Bengtson, gave help and advice during the manuscript preparation. Finally, and most particularly, I thank Hans Christian Bjerring, Docteur ès Sciences, for much useful discussion, for helping organize the material of the paper, and for preparing the final copy.

The publication of this volume has been financially supported by the Swedish Natural Science Research Council.

References

Ahlberg, P.E. 1991: Tetrapod or near-tetrapod fossils from the Upper Devonian of Scotland. *Nature 354,* 298–301.

Ahlberg, P.E., Luksevics, E. & Lebedev, O. 1994: The first tetrapod finds from the Devonian (Upper Famennian) of Latvia. *Philosophical Transactions of the Royal Society of London B 343,* 303–328.

Ahlberg, P.E. & Milner, A.R. 1994: The origin and diversification of tetrapods. *Nature 368,* 507–514.

Balinsky, B.I. 1933: Das Extremitätenseitenfeld; seine Ausdehnung und Beschaffenheit. *Archiv für Entwicklungsmechanik der Organismen 130,* 704–746.

Balinsky, B.I. 1935: Experimentelle Extremitäteninduktion und die Theorien des phylogenetischen Ursprungs der paarigen Extremitäten der Wirbeltiere. *Anatomischer Anzeiger 80,* 136–142.

Balinsky, B.I. 1937: Über die zeitlichen Verhältnisse bei der Extremitäteninduktion. *Archiv für Entwicklungsmechanik der Organismen 136,* 250–285.

Beaumont, E.H. 1977: Cranial morphology of the Loxomatidae (Amphibia: Labyrinthodontia). *Philosophical Transactions of the Royal Society of London B 280,* 29–101.

Beer, G.R. de 1937: *The Development of the Vertebrate Skull.* Oxford University Press, Oxford. 552 pp.

Bellairs, A. d'A. & Kamal, A.M. 1981: The chondrocranium of the skull in Recent reptiles. *In* Gans, C. & Parsons, T.S. (eds.): *Biology of the Reptilia 2 (Morphology),* 1–263. Academic Press, London.

Bendix-Almgreen, S.E., Clack, J.A. & Olsen, H. 1988: Upper Devonian and Upper Permian vertebrates collected in 1987 around Kejser Franz Joseph Fjord, central East Greenland. *Grønlands Geologiske Undersøgelse, Rapport 140,* 95–102.

Bendix-Almgreen, S.E., Clack, J.A. & Olsen, H. 1990: Upper Devonian tetrapod palaeoecology in the light of new discoveries in East Greenland. *Terra Nova 2,* 131–137.

Bergglas, B. 1925: Zur Frage der Hyperdaktylie und des *Os intermetatarsale. Zeitschrift für die Gesamte Anatomie (1. Abt.) 75,* 127–148.

Bjerring, H.C. 1967: Does a homology exist between the basicranial muscle and the polar cartilage? *Colloques Internationaux du Centre National de la Recherche Scientifique 163,* 223–267.

Bjerring, H.C. 1968: The second somite with special reference to the evolution of its myotomic derivatives. *In* Ørvig, T. (ed.): *Current Problems of Lower Vertebrate Phylogeny. Proceedings 4th Nobel Symposium,* 341–357. Almqvist & Wiksell, Stockholm.

Bjerring, H.C. 1970: Nervus tenuis, a hitherto unknown cranial nerve of the fourth metamere. *Acta Zoologica 51,* 107–114. Stockholm.

Bjerring, H.C. 1971: The nerve supply to the second metamere basicranial muscle in osteolepiform vertebrates, with some remarks on the basic composition of the endocranium. *Acta Zoologica 52*, 189–225. Stockholm.

Bjerring, H.C. 1972: Morphological observations on the exoskeletal skull roof of an osteolepiform from the Carboniferous of Scotland. *Acta Zoologica 53*, 73–92. Stockholm.

Bjerring, H.C. 1973: Relationships of coelacanthiforms. *In* Greenwood, P.H., Miles, R.S. & Patterson, C. (eds.): *Interrelationships of Fishes. Zoological Journal of the Linnean Society 53 (Suppl.1)*, 179–205.

Bjerring, H.C. 1977: A contribution to structural analysis of the head of craniate animals. *Zoologica Scripta 6*, 127–183.

Bjerring, H. C. 1978: The 'intracranial joint' *versus* the 'ventral otic fissure'. *Acta Zoologica 59*, 203–214. Stockholm.

Bjerring, H.C. 1984a: The term 'fossa bridgei' and five endocranial fossae in teleostome fishes. *Zoologica Scripta 13*, 231–238.

Bjerring, H.C. 1984b: Major anatomical steps toward craniotedness: a heterodox view based largely on embryological data. *Journal of Vertebrate Paleontology 4*, 17–2 9.

Bjerring, H.C. 1985: Facts and thoughts on piscine phylogeny. *In* Foreman, R.E., Gorbman, A., Dodd, J.M. & Olsson, R. (eds.): *Evolutionary Biology of Primitive Fishes. Nato ASI Series, Ser. A, Life Sciences 103*, 31–57. Plenum Press, New York.

Bjerring. H.C. 1986: Electric tetrapods? *In* Roček, Z. (ed.): *Studies in Herpetology*, 29–36. Proceedings 3rd ordinary general meeting of Societas Europaea Herpetologica. Prague.

Bjerring, H.C. 1988: Arms and legs: an evolutionary perspective. *Fauna och Flora 83*, 58–74. Stockholm. [In Swedish, with an English summary.]

Bjerring, H.C. 1993: A reflection on the evolutionary origin of the tegmen tympani. *Palaeontographica A 228*, 129–142.

Bjerring, H.C. 1995: The parietal problem: how to cut this Gordian knot? *Acta Zoologica 76:3*, 193–203. Stockholm.

Borgen, U. 1983: Homologizations of skull roofing bones between tetrapods and osteolepiform fishes. *Palaeontology 26*, 735–753.

Boy, J.A. 1987: Studien über die Branchiosauridae (Amphibia: Temnospondyli; Ober-Karbon–Unter-Perm). 2. Systematische Übersicht. *Neues Jahrbuch für Geologie und Paläontologie 174*, 75–104.

Boy, J.A. 1988: Über einige Vertreter der Eryopoidea (Amphibia: Temnospondyli) aus dem europäischen Rotliegend (? höchstes Karbon–Perm). 1. *Sclerocephalus. Paläontologische Zeitschrift 62*, 107–132.

Boy, J.A. 1989: Über einige Vertreter der Eryopoidea (Amphibia: Temnospondyli) aus dem europäischen Rotliegend (? höchstes Oberkarbon–Perm). 2. *Acanthostomatops. Paläontologische Zeitschrift 63*, 133–151.

Boy, J.A. 1990: Über einige Vertreter der Eryopoidea (Amphibia: Temnospondyli) aus dem europäischen Rotliegend (? höchstes Karbon–Perm). 3. *Onychiodon. Paläontologische Zeitschrift 64*, 287–312.

Brough, M.C. & Brough, J. 1967: Studies on early tetrapods. 1. The Lower Carboniferous microsaurs. 2. *Microbrachis*, the type microsaur. 3. The genus *Gephyrostegus. Philosophical Transactions of the Royal Society of London B 252*, 107–165.

Bystrow, A.P. 1935: Morphologische Untersuchungen der Deckknochen des Schädels der Wirbeltiere. 1. Schädel der Stegocephalen. *Acta Zoologica 16*, 65–141. Stockholm.

Bystrow, A.P. 1938: *Dvinosaurus* als neotenische Form der Stegocephalen. *Acta Zoologica 19*, 209–295. Stockholm.

Bystrow, A.P. & Efremov, J.A. 1940: *Benthosuchus sushkini* EFR. A labyrinthodont from the Eotriassic of Sharzhenga river. *Trudy Paleontologicheskogo Instituta Akademii Nauk SSSR 10*, 1–152.

Bütler, H. 1935: Die Mächtigkeit der kaledonischen Molasse in Ostgrönland. *Mitteilungen der Naturforschen Gesellschaft Schaffhausen 12*, 17–33.

Bütler, H. 1959: Das Old Red Gebiet am Moskusoksefjord. *Meddelelser om Grønland 160:5*, 1–188.

Bütler, H. 1961: Devonian deposits of central East Greenland. *In* Raasch, G.O. (ed.): *Geology of the Arctic 1*, 188–196. Toronto Press. Toronto, Ont.

Campbell, K.S.W. & Bell, M.W. 1977: A primitive amphibian from the Late Devonian of New South Wales. *Alcheringa 1*, 369–381.

Carroll, R.L. 1967: Labyrinthodonts from the Joggins Formation. *Journal of Paleontology 41*, 111–142.

Carroll, R.L. 1988: *Vertebrate Paleontology and Evolution.* Greeman, New York, N.Y.

Carroll, R.L. 1989: Developmental aspects of lepispondyl vertebrae in Paleozoic tetrapods. *Historical Biology 3*, 1–25.

Carroll, R.L. 1992: The primary radiation of terrestrial vertebrates. *Earth and Planetary Sciences 20*, 45–84.

Case, E.C. 1946: A census of the determinable genera of the Stegocephalia. *Transactions of the American Philosophical Society, New Series 35*, 325–420.

Clack, J.A. 1987: *Pholiderpeton scutigerum* Huxley, an amphibian from the Yorkshire coal measures. *Philosophical Transactions of the Royal Society of London B 318*, 1–107.

Clack, J.A. 1988a: New material of the early tetrapod *Acanthostega* from the Upper Devonian of East Greenland. *Paleontology 31*, 699–724

Clack, J.A. 1988b: Pioneers of the land in East Greenland. *Geodigest: Geology Today*, November–December 1988, 192–194.

Clack, J.A. 1994: *Acanthostega gunnari*, a Devonian tetrapod from Greenland; the snout, palate and ventral parts of the braincase, with a discussion of their significance. *Meddelelser om Grønland, Geoscience 31*, 3–24.

Clack, J.A. & Coates, M.I. 1993: *Acanthostega gunnari*: our present connection. *In* Hoch, E. & Brantsen, A.K. (eds.): *Deciphering the Natural World and the Role of Collections and Museums*, 39–42. Geological Museum, Copenhagen University.

Coates, M.I. 1994: The origin of vertebrate limbs. *Development 1994 Supplement*, 169–180.

Coates, M.I. & Clack, J.A. 1990: Polydactyly in the earliest known tetrapod limbs. *Nature 347*, 66–69.

Daeschler, E. B., Shubin, N. H., Thomson, K. S. & Amaral, W. W. 1994: A Devonian tetrapod from North America. *Science 265*, 639–642.

Dwight, T. 1892: Fusion of hand. *Memoirs of the Boston Society of Natural History 4*, 473–486.

Eeden, J.A. van 1951: The development of the chondrocranium of *Ascaphus truei* Stejneger with special reference to the relations of the palatoquadrate to the neurocranium. *Acta Zoologica, 32*, 41–176. Stockholm.

Friend, P.F., Alexander-Marrack, P.D., Nicholson, J. & Yeats, A.K. 1976: Devonian sediments of East Greenland. 2. Sedimentary structures and fossils. *Meddelelser om Grønland 206:2*, 1–91.

Friend, P.F., Alexander-Marrack, P.D., Allen K. C., Nicholson J. & Yeats A.K. 1983: Devonian sediments of East Greenland: 6. Review of results. *Meddelelser om Grønland 206:6*, 1–96.

Gaupp, E. 1893: Beiträge zur Morphologie des Schädels. 1. Primordial-Cranium und Kieferbogen von *Rana fusca. Morphologische Arbeiten (G. Schwalbe) 2*, 275–481.

Gaupp, E. 1896: Lehre vom Skelet und vom Muskelsystem. *In: A. Ecker's und R. Wiedersheim's Anatomie des Frosches 1*, 1–229.

Gaupp, E. 1899: Lehre vom Nerven und Gefäßsystem. *In: A. Ecker's und R. Wiedersheim's Anatomie des Frosches 2*, 1–548.

Gaupp, E. 1904: Lehre vom Integument und von den Sinnesorganen. *In: A. Ecker's und R. Wiedersheim's Anatomie des Frosches 3:2*, 443–961.

Gaupp, E. 1905: Die Entwickelung des Kopfskelettes. *In* Hertwig, O. (ed.): *Handbuch der vergleichenden und experimentellen Entwickelungslehre der Wirbeltiere 3*, 573–874.

Getmanov, S.N. 1989: Triassic amphibians from the East European platform. *Trudy Paleontologicheskogo Instituta Akademii Nauk SSSR 236*, 1–100. [In Russian.]

Godfrey, S.J. 1989: The postcranial skeletal anatomy of the Carboniferous tetrapod *Greererpeton burkemorani* Romer, 1969. *Philosophical Transactions of the Royal Society of London B 323*, 75–133.

Godfrey, S.J., Forillo, A.R. & Carroll, R.L. 1987: A newly discovered skull of the temnospondyl amphibian *Dendrerpeton acadianum* Owen. *Canadian Journal of Earth Sciences 24*, 796–805.

Goodrich, E.S. 1930: *Studies on the Structure and Development of Vertebrates.* 837 pp. Macmillan, London.

Gould, S.J. 1991: *Eight (or fewer) little piggies. Natural History 1*, 22–29. New York, N.Y.

Gould, S.J. 1993: *Eight Little Piggies.* 479 pp. Jonathan Cape, London.

Gray, H. 1973: *Anatomy of the Human Body* (ed. C.M. Goss). 29th American Ed. Lea & Febiger, Philadelphia, Penn.

Gregory, W.K. & Raven, H.C. 1941: Studies on the origin and early evolution of paired fins and limbs. *Annals of the New York Academy of Sciences 42*, 273–360.

Gross, W. 1941: Über den Unterkiefer einiger devonischer Crossopterygier. *Abhandlungen der Preussischen Akademie der Wissenschaften* 7, 3–51.

Gross, W. 1956: Über Crossopterygier und Dipnoer aus dem baltischen Oberdevon im Zusammenhang einer vergleichenden Untersuchung des Porenkanalsystems paläozoischer Agnathen und Fische. *Kungliga Svenska Vetenskapsakademiens Handlingar (4)* 5, 1–140.

Guibé, J. 1970: Le squelette du tronc et des membres. *In*: Grassé, P.-P.(ed.): *Traité de Zoologie 14*, 33–77.

Heintz, A. 1930: Oberdevonische Fischreste aus Ost-Grönland. *Skrifter om Svalbard og Ishavet 30*, 30–46.

Heintz, A. 1932: Beitrag zur Kenntnis der devonischen Fischfauna Ost-Grönlands. *Skrifter om Svalbard och Ishavet 42*, 5–27.

Hinchliffe, J.R. 1977: The chondrogenic pattern in chick limb morphogenesis: A problem of development and evolution. *In*: Ede, D. A., Hinchliffe, J.R. & Balls, M. (eds.): *Vertebrate Limb and Somite Morphogenesis*, 293–309. Cambridge University Press, Cambridge.

Hinchliffe, J.R. & Johnson, D.R. 1980: *The Development of the Vertebrate Limb*. 268 pp. Clarendon, Oxford.

Holmes, R. 1980: *Proterogyrinus scheelei* and the early evolution of the labyrinthodont pectoral limb. *In* Panchen, A.L. (ed.): *The Terrestrial Environment and the Origin of Land Vertebrates*, 351–376.

Holmes, R. 1989: The skull and axial skeleton of the Lower Permian anthracosaurid amphibian *Archeria crassidisca* Cope. *Palaeontographica 207*, 161–206.

Holmgren, N. 1940: Studies on the head in fishes. Embroylogical, morphological, and phylogenetical researches. 1. Development of the skull in sharks and rays. *Acta Zoologica 21*, 51–267. Stockholm.

Holmgren, N. 1941: Studies on the head in fishes. Embryological, morphological, and phylogetical researches. 2. Comparative anatomy of the adult selachian skull, with remarks on the dorsal fins in sharks. *Acta Zoologica 22*, 1–100. Stockholm.

Holmgren, N. 1942: Studies on the head of fishes. An embryological, morphological, and phylogenetical study. 3. The phylogeny of elasmobranch fishes. *Acta Zoologica 23*, 129–261. Stockholm.

Holmgren, N. 1943: Studies on the head of fishes. An embryological, morphological, and phylogenetical study. 4. General morphology of the head in fish. *Acta Zoologica 24*, 1–188. Stockholm.

Hook, R.W. 1983: *Colosteus scutellatus* (Newberry), a primitive temnospondyl amphibian from the Middle Pennsylvanian of Linton, Ohio. *American Museum Novitates 270*, 1–41.

Hurle, J.M. & Ganan, Y. 1986: Interdigital tissue chondrogenesis induced by surgical removal of the ectoderm in the embryonic chick leg bud. *Journal of Embryology and Experimental Morphology 94*, 231–244.

Jarvik, E. (Johansson, A.E.V.) 1935: Upper Devonian fossiliferous localities in Parallel valley on Gauss Peninsula, East Greenland investigated in the summer of 1934. *Meddelser om Grønland 96:3*, 1–37.

Jarvik, E. 1937: On the species of *Eusthenopteron* found in Russia and the Baltic states. *Bulletin of the Geological Institution of the University of Upsala 27*, 63–127.

Jarvik, E. 1942: On the structure of the snout of crossopterygians and lower gnathostomes in general. *Zoologiska Bidrag från Uppsala 21*, 235–675.

Jarvik, E. 1944: On the dermal bones, sensory canals and pit-lines of the skull in *Eusthenopteron foordi* Whiteaves, with some remarks on E. *saeve-soederberghi* Jarvik. *Kungliga Svenska Vetenskapsakademiens Handlingar 21*, 1–48.

Jarvik, E. 1948a: On the morphology and taxonomy of the Middle Devonian osteolepid fishes of Scotland. *Kungliga Svenska Vetenskapsakademiens Handlingar 25*, 1–301.

Jarvik, E. 1948b: Note on the Upper Devonian vertebrate fauna of East Greenland and on the age of the ichthyostegid stegocephalians. *Arkiv för Zoologi 41A*, 1–8.

Jarvik, E. 1950a: Middle Devonian vertebrates from Canning Land and Wegeners Halvø (East Greenland) 2. Crossopterygii. *Meddelelser om Grønland 96:4*, 1–132.

Jarvik, E. 1950b: Note on Middle Devonian crossopterygians from the eastern part of Gauss Halvø, East Greenland. With an appendix: An attempt at a correlation of the Upper Old Red Sandstone of East Greenland with the marine sequence. *Meddelelser om Grønland 149:6*, 1–20.

Jarvik, E. 1952: On the fish-like tail in the ichthyostegid stegocephalians with descriptions of a new stegocephalian and a new crossopterygian from the Upper Devonian of East Greenland. *Meddelelser om Grønland 114(12)*, 1–90.

Jarvik, E. 1954: On the visceral skeleton in *Eusthenopteron* with a discussion of the parasphenoid and palatoquadrate in fishes. *Kungliga Svenska Vetenskapsakademiens Handlingar (4)* 5, 1–104.

Jarvik, E. 1955a: The oldest tetrapods and their forerunners. *Scientific Monthly 80*, 141–154.

Jarvik, E. 1955b: Ichthyostegalia. *In* J. Piveteau (ed.):*Traité de Paléontologie 5*, 53–66. Masson, Paris.

Jarvik, E. 1959a: Dermal fin-rays and Holmgren's principle of delamination. *Kungliga Svenska Vetenskapsakademiens Handlingar (4)* 6, 1–51.

Jarvik, E. 1959b: De tidiga fossila ryggradsdjuren. Paleoanatomiska arbetsmetoder och resultat. [The early fossil vertebrates. Paleoanatomical methods and results.] *Svensk Naturvetenskap 1959*, 5–80. Stockholm. [In Swedish.]

Jarvik, E. 1960: *Théories de l'évolution de Vertébrés reconsidérées à la Lumiere des récentes Découvertes sur les Vertébrés inférieurs.* Préface et traduction de J.-P. Lehman, 3–104. Masson, Paris.

Jarvik, E. 1961: Devonian vertebrates. *In* Raasch, G.O. (ed.): *Geology of the Arctic 1*, 197–204. University of Toronto Press, Toronto, Ont.

Jarvik, E. 1962: Les porolépiformes et l'origine des urodèles. *Colloques Internationales du Centre National de Recherche Scientifique 104*, 87–101.

Jarvik, E. 1963: The composition of the intermandibular division of the head in fish and tetrapods and the diphyletic origin of the tetrapod tongue. *Kungliga Svenska Vetenskapsakademiens Handlingar (4)* 9, 1–74.

Jarvik, E. 1964: Specializations in early vertebrates. *Annales de la Société Royale Zoologique de Belgique 94*, 1–95.

Jarvik, E. 1965a: Die Raspelzunge der Cyclostomen und die pentadactyle Extremität der Tetrapoden als Beweise für monophyletische Herkunft. *Zoologischer Anzeiger 175*, 101–143.

Jarvik, E. 1965b: On the origin of girdles and paired fins. *Israel Journal of Zoology 14*, 141–172.

Jarvik, E. 1966: Remarks on the structure of the snout in *Megalichthys* and certain other rhipidistid crossopterygians. *Arkiv för Zoologi 19*, 41–98.

Jarvik, E. 1967: The homologies of frontal and parietal bones in fishes and tetrapods. *Colloques internationaux du Centre national de la Recherche scientifique 163*, 181–213.

Jarvik, E. 1968: Aspects of vertebrate phylogeny. *In* T. Ørvig (ed.): *Current Problems of Lower Vertebrate Phylogeny. Nobel Symposium 4*, 497–527. Almqvist & Wiksell, Stockholm.

Jarvik, E. 1972: Middle and Upper Devonian Porolepiformes from East Greenland with special references to *Glyptolepis groenlandica* n.sp. *Meddelelser om Grønland 187:2*, 1–295.

Jarvik, E. 1975: On the *saccus endolymphaticus* and adjacent structures in osteolepiforms, anurans and urodeles. *Colloques internationaux du Centre National de la Recherche scientifique 218*, 191–211.

Jarvik, E. 1980a: *Basic Structure and Evolution of Vertebrates 1.* 575 pp. Academic Press, London.

Jarvik, E. 1980b: *Basic Structure and Evolution of Vertebrates 2.* 337 pp. Academic Press, London.

Jarvik, E. 1981: [Review of Rosen, D.E., Forey, P.L., Gardiner, B.G. & Patterson, C.: Lungfishes, Tetrapods, Paleontology, and Plesiomorphy.] *Systematic Zoology 30*, 378–384.

Jarvik, E. 1985: Devonian osteolepiform fishes from East Greenland. *Meddelelser om Grønland, Geoscience 13*, 1–52.

Jarvik, E. 1986: On the origin of the Amphibia. *In* Roček, Z. (ed.): *Studies in Herpetology.* Proceedings of the 3rd Ordinary General Meeting of Societas Europaea Herpetologica, Prague, 1–24.

Jarvik, E. 1988: The early vertebrates and their forerunners. *In*: *L'évolution dans sa Réalité et ses diverses Modalités*, 35–64. Fondation Singer-Polignac & Masson, Paris.

Jessen, H. 1966: Die Crossopterygier des Oberen Plattenkalkes (Devon) der Bergisch-Gladbach-Paffrather Mulde (Rheinisches Schiefergebirge) unter Berücksichtigung von amerikanischem und europäischem *Onychodus*-Material. *Arkiv för Zoologi 128*, 87–114.

Jupp, R. & Warren, A. 1986: The mandibles of Triassic temnospondyl amphibians. *Alcheringa 10*, 99–124.

Klembara, J. 1991: Nové poznatky o najstarších štvornožcoch. *Vesmír 70*, 88–91. Praha.

Klembara, J. 1992: The first record of pit-lines and foraminal pits in tetrapods and the problem of the skull roof bones between tetrapods and fishes. *Geologica Carpathica 42*, 249–252. Bratislava.

Klembara, J. 1993: The subdivisions and fusions of the exoskeletal skull bones of *Discosauriscus austriacus* (Makowsky 1876) and their possible homologues in rhipidistians. *Paläontologische Zeitschrift 67*, 145–168.

Koch, L. 1930: Preliminary report on the Danish expedition to East Greenland in 1929. *Meddelelser om Grønland 74*, 173–206.

Koch, L. & Haller, J. 1971. Geological map of East Greenland: 72°–76°N. Lat. (1:250000). *Meddelelser om Grønland 183*, 1–26.

Kulling, O. 1930: Stratigraphic studies of the geology of Northeast Greenland. *Meddelelser om Grønland 74*, 317–346.

Kulling, O. 1931: An account of the localities of the Upper Devonian vertebrate finds in East Greenland in 1929, *Meddelelser om Grønland 86:2*, 1–14.

Lauder, G.V. 1980: On the relationship of the myotome to the axial skeleton in vertebrate evolution. *Paleobiology 6*, 51–56.

Lebedev, A.O. 1984: The first find of a Devonian tetrapod vertebrate in the U.S.S.R. *Doklady Akademii Nauk SSSR, Palaeontology 278*, 1470–1473.

Lebedev, O.A. 1985: The first tetrapods: searches and findings. *Priroda 11*, 26–36 (In Russian).

Lebedev, A.O. 1990: *Tulerpeton*, l'animal a six doigts. *La Recherche 225*, 1274–1275

Lebedev, O.A.. & Clack, J.A. 1993: Upper Devonian tetrapods from Andreyevka, Tula Region, Russia. *Palaeontology 36*, 721–734.

Lehman, J.-P. 1959: Les Dipneustes du Dévonien supérieur du Groenland. *Meddelelser om Grønland 164:4*, 1–58.

Leonardi, G. 1983: *Notopus petri* nov.gen, nov.sp: Une Empreinte d'amphibien du Dévonian au Parana (Brésil). *Geobios 16*, 233–239.

Long, J.A. 1985a: New information on the head and shoulder girdle of *Canowindra grossi* Thomson, from the Late Devonian Mandagery Sandstone, New South Wales. *Records of the Australian Museum 37*, 91–99.

Long, J.A. 1985b: The structure and relationships of a new osteolepiform fish from the Late Devonian of Victoria, Australia. *Alcheringa 9*, 1–22.

Long, J.A. 1987: An unusual osteolepiform fish from the Late Devonian of Victoria, Australia. *Palaeontology 30*, 839–852.

Milner, A.R. 1980: The temnospondyl amphibian *Dendrerpeton* from the Upper Carboniferous of Ireland. *Palaeontology 23*, 125–141.

Miner, R.W. 1925: The pectoral limb of *Eryops* and other primitive tetrapods. *Bulletin of the American Museum of Natural History 51*, 145–312.

Mook, C.C. 1921: The dermo-supraoccipital bone in the Crocodilia. *Bulletin of the American Museum of Natural History 44*, 101–103.

Müller, E. 1909: Die Brustflosse der Selachier. *Arbeiten aus Anatomischen Instituten in Wiesbaden 39*, 469–601.

Müller, E. 1911: Untersuchungen über die Muskeln und Nerven der Brustflosse und der Körperwand bei *Acanthias vulgaris*. *Arbeiten aus Anatomischen Instituten in Wiesbaden 43*, 1–147.

Nathorst, A.G. 1900: Den svenska expeditionen till nordöstra Grönland 1899. *Ymer 1900 (häfte 2)*, 115–156. Svenska Sällskapet för Antropologi och Geografi, Stockholm. [In Swedish.]

Nathorst, A.G. 1901: Bidrag till nordöstra Grönlands geologi. *Geologiska Föreningens i Stockholm Förhandlingar 23*, 275–306. [In Swedish.]

Nilsson, T. 1939: Cleithrum und Humerus der Stegocephalen und rezenten Amphibien auf Grund neuer Funde von *Plagiosaurus depressus* Jaekel. *Kungliga Fysiografiska Sällskapets i Lund Handlingar 50*, 1–39.

Nilsson, T. 1942: *Sassenisaurus*, a new genus of Eotriassic stegocephalians from Spitsbergen. *Bulletin of the Geological Institution of the University of Upsala 30*, 91–102.

Nilsson, T. 1943: On the morphology of the lower jaw of Stegocephalia, with special reference to Eotriassic stegocephalians from Spitsbergen. 1. Descriptive part. *Kungliga Svenska Vetenskapsakademiens Handlingar 20*, 1–46.

Nilsson, T. 1944: On the morphology of the lower jaw of Stegocephalia with special reference to Eotriassic stegocephalians from Spitsbergen 2. General part. *Kungliga Svenska Vetenskapsakademiens Handlingar 21*, 1–70.

Nilsson, T. 1946: On the genus *Peltostega* Wiman and the classification of the Triassic stegocephalians. *Kungliga Svenska Vetenskapsakademiens Handlingar 23*, 1–55.

Olsen, H. 1993: Sedimentary basin analysis of the continental Devonian basin in North-East Greenland. *Bulletin, Grønlands Geologiske Undersøgelse 169*, 1–80.

Olsen, H. & Larsen, P.-H. 1993: Lithostratigraphy of the continental Devonian sediments in North-East Greenland. *Bulletin, Grønlands Geologiske Undersøgelse 165*, 1–108.

Orvin, A.K. 1930: Beiträge zur Kenntnis des Oberdevon Ost-Grönlands. *Skrifter om Svalbard og Ishavet 30*, 1–46.

Panchen, A.L. 1964: The cranial anatomy of two Coal Measure anthracosaurs. *Philosophical Transactions of the Royal Society of London B 247*, 593–637.

Panchen, A.L. 1972: The skull and skeleton of *Eogyrinus attheyi* Watson (Amphibia:Labyrinthodontia). *Philosophical Transactions of the Royal Society of London B 263*, 279–326.

Panchen, A.L. 1975: A new genus and species of anthracosaur amphibian from the Lower Carboniferous of Scotland and the status of *Pholidogaster pisciformis* Huxley. *Philosophical Transactions of the Royal Society of London B 269*, 581–640.

Panchen, A.L. 1977: On *Anthracosaurus russelli* Huxley (Amphibia:Labyrinthodontia) and the family Anthracosauridae. *Philosophical Transactions of the Royal Society of London B 279*, 447–512.

Panchen, A.L. 1985: On the amphibian *Crassigyrinus scoticus* Watson from the Carboniferous of Scotland. *Philosophical Transactions of the Royal Society of London B 309*, 505–568.

Panchen, A.L. & Smithson, T.R. 1987: Character diagnosis, fossils and the origin of tetrapods. *Biological Reviews 62*, 341–438. Cambridge.

Panchen, A.L. & Smithson, T.R. 1990: The pelvic girdle and hind limb of *Crassigyrinus scoticus* (Lydekker) from the Scottish Carboniferous and the origin of the tetrapod pelvic skeleton. *Transactions of the Royal Society of Edinburgh, Earth Sciences 81*, 31–44.

Paterson, N.F. 1949: The development of the inner ear of *Xenopus laevis*. *Proceedings of the Zoological Society of London 119*, 269–291.

Regel, E. D. 1961. Traces of segmentation in the chordal division of the chondrocranium of *Hynobius kayserlingii*. *Doklady Akademii Nauk SSSR 140*, 253–255. [In Russian, with English title.]

Regel, E. D. 1964. The development of the cartilaginous neurocranium and its connection with the palatoquadrate in *Hynobius keyserlingii*. *Trudy Zoologicheskogo Instituta Akademii Nauk SSSR, Leningrad 33*, 34–74. [In Russian, with English title.]

Regel, E. D. 1968. The development of the cartilaginous neurocranium and its connection with the upper part of the mandibular arch in the Sibirian salamander *Ranodon sibiricus* (Hynobiidae, Amphibia). *Trudy Zoologicheskogo Instituta Akademii Nauk SSSR, Leningrad 46*, 5–85. [In Russian, with English title.]

Roček, Z. 1985: Tooth replacement in *Eusthenopteron* and *Ichthyostega*. *Fortschritte der Zoologie 30*, 249–252.

Roček, Z. & Rage, J.-C. 1994: The presumed amphibian footprint *Notopus petri* from the Devonian: a probably starfish trace fossil. *Lethaia 27:3*, 241–244.

Romer, A. S. 1930: The Pennsylvanian tetrapods of Linton, Ohio. *Bulletin of the American Museum of Natural History 59*, 77–147.

Romer, A.S. 1936a: The dipnoan cranial roof. *American Journal of Science 32*, 242–256.

Romer, A.S. 1936b; Review of: G. Säve-Söderbergh, 1935a, On the dermal bones of the head in labyrinthodont stegocephalians and primitive reptilia. *Journal of Geology 44*, 534–536.

Romer, A.S. 1937: The braincase of the Carboniferous crossopterygian *Megalichthys nitidus*. *Bulletin of the Museum of Comparative Zoology at Harvard 82*, 1–73.

Romer, A.S. 1941: Notes on the crossopterygian hyomandibular and braincase. *Journal of Morphology 69*, 141–160.

Romer, A.S. 1944: The development of tetrapod limb musculature. The shoulder region of *Lacerta*. *Journal of Morphology 74*, 1–41.

Romer, A.S. 1946: The early evolution of fishes. *Quarterly Review of Biology 21*, 33–69.

Romer, A.S. 1947: Review of the Labyrinthodontia. *Bulletin of the Museum of Comparative Zoology at Harvard 99*, 1–368.

Romer, A.S. 1956a: The early evolution of land vertebrates. *Proceedings of the American Philosophical Society 100*, 157–167.

Romer, A. S. 1956b: *Osteology of the Reptiles.* University Press, Chicago. 772 pp.

Romer, A.S. 1957: The appendicular skeleton of the Permian embolomerous amphibian *Archeria. Contributions from the Museum of Paleontology, University of Michigan 13*, 103–159.

Romer, A.S. 1962: [Review of J.-P. Lehman *L'Évolution des Vertébrés Inférieurs* and E. Jarvik *Théories de l'Évolution des Vertébrés.*] *Copeia 1962*, 223–227.

Romer, A.S. 1966: *Vertebrate Paleontology.* 3rd Ed. 468 pp. The University of Chicago Press, Chicago, Ill.

Romer, A.S. 1970: *The Vertebrate Body.* 4th Ed. 601 pp. Saunders Company, Philadelphia, Pa.

Romer, A.S. & Witter, R.V. 1942: *Edops*, a primitive rhachitomous amphibian from the Texas Red Beds. *Journal of Geology 50*, 925–960.

Säve-Söderbergh, G. 1932a: Preliminary note on Devonian stegocephalians from East Greenland. *Meddelelser om Grønland 94:7*, 1–107.

Säve-Söderbergh, G. 1932b: Notes on Devonian stratigraphy of East Greenland. *Meddelelser om Grønland 94:4*, 1–40.

Säve-Söderbergh, G. 1933a: The dermal bones of the head and the lateral line system in *Osteolepis macrolepidotus* Ag. With remarks on the terminology of the lateral line system and on the dermal bones of certain other crossopterygians. *Nova Acta Regiae Societatis Scientiarum Upsaliensis (4) 9*, 1–129.

Säve-Söderbergh, G. 1933b: Further contributions to the Devonian stratigraphy of East Greenland. 1. Results from the summer expedition 1932. *Meddelelser om Grønland 96:1*, 1–40.

Säve-Söderbergh, G. 1934a: Some points of view concerning the evolution of the vertebrates and the classification of this group. *Arkiv för Zoologi 26A*, 1–20.

Säve-Söderbergh, G. 1934b: Further contributions to the Devonian stratigraphy of East Greenland. 2. Investigations on Gauss Peninsula during the summer of 1933. With an appendix: Notes on the geology of the Passage Hills (East Greenland). *Meddelelser om Grønland 96:2*, 1–74.

Säve-Söderbergh, G. 1935a: On the dermal bones of the head in labyrinthodont stegocephalians and primitive Reptilia with special reference to Eotriassic stegocephalians from East Greenland. *Meddelelser om Grønland 98:3*, 1–211.

[Säve-Söderbergh, G. 1935b: Further contributions to the Devonian stratigraphy of East Greenland. 3. Investigations in the Upper Devonian of the Franz Joseph Fjord district in 1934. 27 pp. Manuscript in the Department of Palaeozoology, Swedish Museum of Natural History, Stockholm.]

Säve-Söderbergh, G. 1936: On the morphology of Triassic stegocephalians from Spitsbergen and the interpretation of the endocranium in the Labyrinthodontia. *Kungliga Svenska Vetenskapsakademiens Handlingar (3) 16*, 1–181.

[Säve-Söderbergh, G. 1937. Further contributions to the Devonian stratigraphy of East Greenland. 4. Stratigraphy of the Parallel Valley area on Gauss Peninsula. Investigations in the Upper Devonian of the Franz Joseph Fjord district in 1936. 26 pp. Manuscript in the Department of Palaeozoology, Swedish Museum of Natural History, Stockholm.]

[Säve-Söderbergh, G. 1938: Further contributions to the Devonian stratigraphy of East Greenland. 5. Notes on the geology of Gästis Valley and the age of the Upper Sandstone Complex. 18 pp. Manuscript in the Department of Palaeozoology, Swedish Museum of Natural History, Stockholm.]

Säve-Söderbergh, G. 1941: On the dermal bones of the head in *Osteolepis macrolepidotus* Ag. and the interpretation of the lateral line system in certain primitive vertebrates. *Zoologiska Bidrag från Uppsala 20*, 523–541.

Säve-Söderbergh, G. 1945; Notes on the trigeminal musculature in non-mammalian tetrapods. *Nova Acta Regiae Societatis Scientiarum Upsaliensis (4) 13*, 1–59.

Säve-Söderbergh, G. 1946: Drag ur den svenska vertebratpaleontologiens utveckling och resultat. *Geologiska Föreningens i Stockholm Förhandlingar 68*, 352–371. [In Swedish.]

Schultze, H.-P. 1969: Die Faltenzähne der Rhipidistiden Crossopterygier, der Tetrapoden und der Actinopterygier-Gattung *Lepisosteus*; Nebst einer Beschreibung der Zahnstruktur von *Onychodus. Palaeontographia Italica 65*, 63–136.

Schultze, H.-P. & Arsenault, M. 1985: The panderichthyid fish *Elpistostege*: a close relative of tetrapods? *Palaeontology 28:2*, 293–309.

Schultze, H.-P. & Arratia, G. 1986: Reevaluation of the caudal skeleton of actinopterygian fishes: 1. *Lepisosteus* and *Amia. Journal of Morphology 190*, 215–241.

Schultze, J.-P. & Arsenault, M. 1987: *Quebecius quebecensis* (Whiteaves) from the Late Devonian of Quebec, Canada. *Canadian Journal of Earth Sciences 24*, 2351–2361.

Sewertzoff, A.N. 1904: Die Entwickelung der pentadaktylen Extremität der Wirbeltiere. *Anatomischer Anzeiger 25*, 472–494.

Sewertzoff, A.N. 1908: Studien über die Entwickelung der Muskeln, Nerven und des Skeletts der Extremitäten der niederen Tetrapod. *Byulleten' Moskovskogo Obshchestva Ispytatelei Prirody 21*, 1–430.

Sewertzoff, A.N. 1926a: Die Morphologie der Brustflossen der Fische. *Jenaische Zeitschrift für Naturwissenschaft 62*, 343–392.

Sewertzoff, A.N. 1926b: Development of the pelvic fins of *Acipenser ruthenus*. New data for the theory of the paired fins of fishes. *Journal of Morphology 41*, 547–579.

Sewertzoff, A.N. 1934: Evolution der Bauchflossen der Fische. *Zoologischer Jahrbücher 58*, 415–500.

Shishkin, M.A. 1973: The morphology of the early Amphibia and some problems of the lower tetrapod evolution. *Trudy Paleontologicheskogo instituta. Akademiya nauk SSSR 137*, 1–260 (In Russian).

Shishkin, M.A. 1989: The axial skeleton of early amphibians and the origin of resegmentation in tetrapod vertebrae. *Fortschritt der Zoologie/Progress in Zoology 35*, 180–195.

Shubin, N.H. & Alberch, P. 1986: A morphogenetic approach to the origin and basic organization of the tetrapod limb. *In* Hecht, M.K., Wallace, B. & France, G.T. (eds.): *Evolutionary Biology 20*, 319–387. Plenum, New York

Smit, A.L. 1953: The ontogenesis of the vertebral column of *Xenopus laevis* (Daudin) with special reference to the segmentation of the metotic region of the skull. *Annals of the University of Stellenbosch 29A:3*, 79–136.

Smithson, T.R. 1982: The cranial morphology of *Greererpeton burkemorani* Romer (Amphibia: Temnospondyli). *Zoological Journal of the Linnean Society 76*, 29–90.

Smithson, T.R. 1985: The morphology and relationships of the Carboniferous amphibian *Eoherpeton watsoni* Panchen. *Zoological Journal of the Linnean Society 85*, 317–410.

Sobotta, J. 1922: *Descriptive Anatomie.* 1. Knochen, Bänder, Belenke, Regionen und Muskeln des menschlichen Körpers. *Lehmanns Medizinische Atlanten II*, 1–263. Lehmann, München.

Špinar, Z.V. 1972: *Tertiary Frogs from Central Europe.* 286 pp. Academia Publishing House of the Czechoslovak Academie of Sciences. Prague.

Spjeldnæs, N. 1982: Palaeoecology of *Ichthyostega* and the origin of the terrestrial vertebrates. *In* Gallitelli, E.M. (ed.): *Palaeontology, Essential of Historical Geology.* Proceedings of the 1st international meeting on 'Palaeontology, Essential of Historical Geology'. Venice 1981, 323–343.

Stadtmüller, F. 1936: Kranium und Visceralskelett der Stegocephalen und Amphibien. *In* Bolk, L. *et al.* (eds.): *Handbuch der vergleichenden Anatomie der Wirbeltiere 4*, 501–698. Urban & Schwarzenberg, Berlin & Wien.

Starck, D. 1975: *Embryology.* 3rd Ed., Thieme, Stuttgart. 704 pp.

Stensiö, E. 1931: Upper Devonian vertebrates from East Greenland. *Meddelelser om Grønland 86:1*, 1–212.

Stensiö, E. 1934: On the Placodermi of the Upper Devonian of East Greenland. I. Phyllolepida and Arthrodira. *Meddelelser om Grønland 97:1*, 1–58.

Stensiö, E. 1936: On the Placodermi of the Upper Devonian of East Greenland. Supplement to part I. *Meddelelser om Grønland 97:2*, 1–52.

Stensiö, E. 1939: On the Placodermi of the Upper Devonian of East Greenland. Second supplement to part I. *Meddelelser om Grønland 97:3*, 1–33.

Stensiö, E. 1947: Sensory lines and dermal bones of the cheek in fishes and amphibians. *Kungliga Svenska Vetenskapsakademiens Handlingar (3) 24*, 1–195.

Stensiö, E. 1948: On the Placodermi of the Upper Devonian of East Greenland. 2. Antiarchi: subfamily Bothriolepinae. *Meddelelser om Grønland 139:1*, (*Palaeozoologica Groenlandica 2*), 5–622.

Stensiö, E. 1959: On the pectoral fin and shoulder girdle of the arthrodires. *Kungliga Svenska Vetenskapsakademiens Handlingar (4) 8:1*, 5–226.

Swanepoel, J.H. 1970: The ontogenesis of the chondrocranium and of the nasal sac of the microhylid frog *Breviceps adspersus pentheri* Werner. *Annale Universiteit van Stellenbosch 45*, 1–119.

Tabin, C.J. 1992: Why we have (only) five fingers per hand: Hox genes and the evolution of paired limbs. *Development 116*, 289–296.

Tatarinov, L.P. 1994: On unusual peculiarities of head morphology in therocephalian *Hexacynodon purlinensis*. *Russian Journal of Herpetology 1*, 1–12.

Vorobyeva, E. I. 1962: Rhizodont crossopterygians from the Devonian Main Field of the USSR. *Trudy Paleontologicheskogo Instituta Akademii Nauk SSSR 104*, 1–108. [In Russian.]

Vorobyeva, E.I. 1975: Bemerkungen zu *Panderichthys rhombolepis* (Gross) aus Lode in Lettland (Gauja Schichten, Oberdevon). *Neues Jahrbuch für Geologie und Paläontologie, Monatshefte 1975:5*, 315–320.

Vorobyeva, E.I. 1977: Morphology and nature of evolution of crossopterygian fish. *Trudy Paleontologicheskogo Instituta Akademii Nauk SSSR 163*, 1–239. [In Russian.]

Vorobyeva, E.I. 1980: Observations on two rhipidistian fishes from the Upper Devonian of Lode, Latvia. *Zoological Journal of the Linnean Society 70*, 191–201.

Vorobyeva, E.I. & Lebedev, O.A. 1986: *Peregrinia krasnovi*, a new species of Glyptopominae (Crossopterygii). *Paleontologicheskii zhurnal 1986:3*, 123–126.

Wake, D.B. & Wake, M.H. 1986: On the development of vertebrae in gymnophione amphibians. *Mémoires de la Socété Zoologique de France 43*, 67–70.

Warren, A., Jupp, R. & Bolton, B. 1986: Earliest tetrapod trackway. *Alcheringa 10*, 183–186.

Warren, A. & Snell, N. 1991: The postcranial skeleton of Mesozoic temnospondyl amphibians: a review. *Alcheringa, 15*, 43–64.

Warren, J.W. & Wakefield, N.A. 1972: Trackways of tetrapod vertebrates from the Upper Devonian of Victoria, Australia. *Nature 238*, 469–470.

Watson, D.M.S. 1926: The evolution and origin of the Amphibia. *Philosophical Transactions of the Royal Society of London B 214*, 189–257.

Watson, D.M.S. 1962: The evolution of the labyrinthodonts. *Philosophical Transactions of the Royal Society of London B 245*, 219–265.

Welles, S.P. & Cosgriff, J. 1965: A revision of the labyrinthodont family Capitosaurida. *University of California Publications in Geological Sciences 54*, 1–148.

Westhuizen, C.M. van der 1961: The development of the chondrocranium of *Heleophryne purcelli* Sclater with special reference to the palatoquadrate and the sound-conducting apparatus. *Acta Zoologica 42*, 3–72. Stockholm.

Westoll, T.S. 1936: On the structures of the dermal ethmoid shield of *Osteolepis*. *Geological Magazine 73*, 151–171.

Westoll, T.S. 1937: The Old Red Sandstone fishes of the north of Scotland particularly of Orkney and Shetland. *Geologists Association Proceedings 48*, 13–45.

Westoll, T.S. 1938: Ancestry of the tetrapods. *Nature 141*, 127–128.

Westoll, T.S. 1940: New Scottish material of *Eusthenopteron*. *Geological Magazine 77*, 65–73.

Westoll, T.S. 1943: The origin of the tetrapods. *Biological Reviews 18*, 78–98.

Wiman, C. 1916: Neue Stegocephalenfunde aus dem Posidonomya Schiefer Spitzbergens. *Bulletin of the Geological Institution of the University of Upsala 13:2*, 209–222.

Woodward, A.S. 1900: Notes on some Upper Devonian fish-remains discovered by Professor A.G. Nathorst in East Greenland. *Bihang till Kungliga Svenska Vetenskapsakademiens Handlingar 26:4*, 1–10.

Young, G.C., Long, J.A. & Ritchie, A. 1992: Crossopterygian fishes from the Devonian of Antarctica: Systematics, relationships and biogeographic significance. *Records of the Australian Museum, Supplement 14*, 1–77.

Plates

Photographs of specimens of *Ichthyostega* collected by the Danish expeditions to East Greenland 1929–1955. All specimens belong to the Geological Museum of the University of Copenhagen (see p. 19 regarding their catalogue numbers).

Plate 1

☐1, 2. Imperfect skull in dorsal and lateral views. Specimen no.220, coll. 1929, Celsius Bjerg. Natural size.

☐3. Imperfect skull in lateral view. A. 245, coll. 1955, Celsius Bjerg. Natural size.

fe.exch, fenestra exochoanalis; *rect.inc*, rectangular incisure.

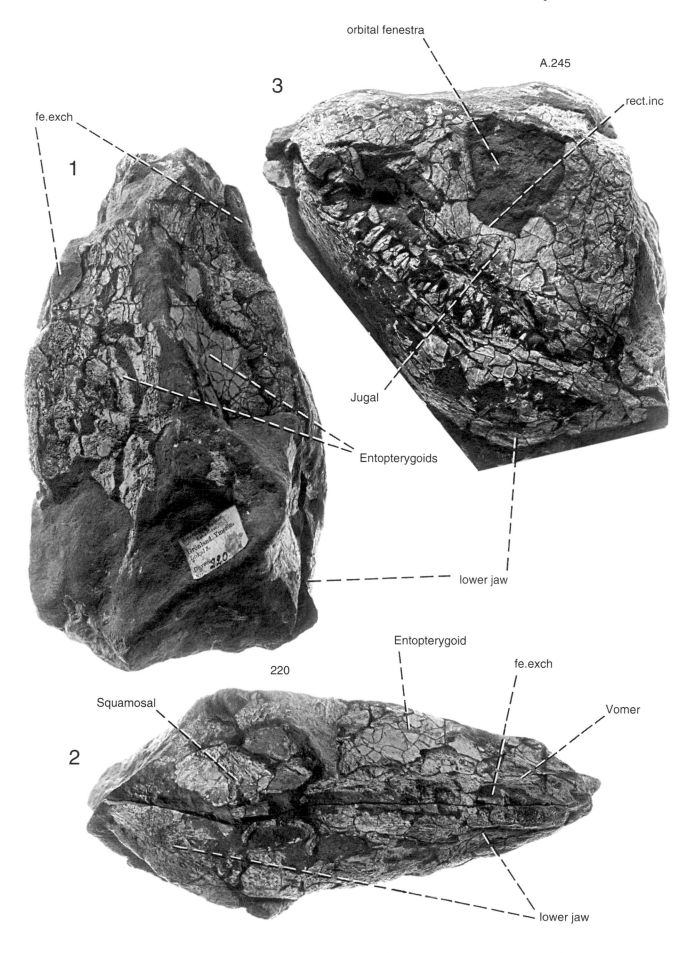

FOSSILS AND STRATA 40 (1996)

Plate 2

Latex casts of two skull tables in dorsal view. $\times \frac{2}{3}$.

☐1. A. 71, coll. 1936, Sederholm Bjerg.

☐2. A. 64, coll. 1934, Remigolepisryg (same specimen as in Pls. 3, 4, 10).

ar.sn.li, area for attachment of supraneural ligament; *c.a.occ*, canal for occipital artery; *ioc*, groove for infraorbital sensory canal; *od.Pop*, area overlapped by preopercular; *vert.la*, vertical complex lamina.

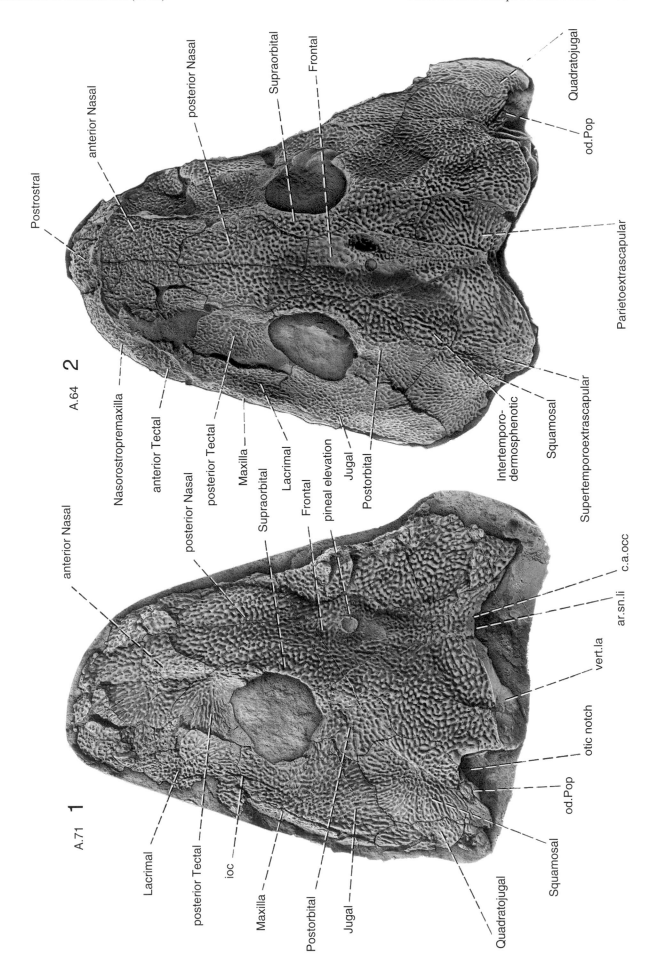

Plate 3

Skull in dorsal view. A. 64. Cast of counterpart in Pl. 2:2.
Natural size.

ar.sn.li, area for attachment of supraneural ligament; *c.a.occ*,
canals for occipital arteries.

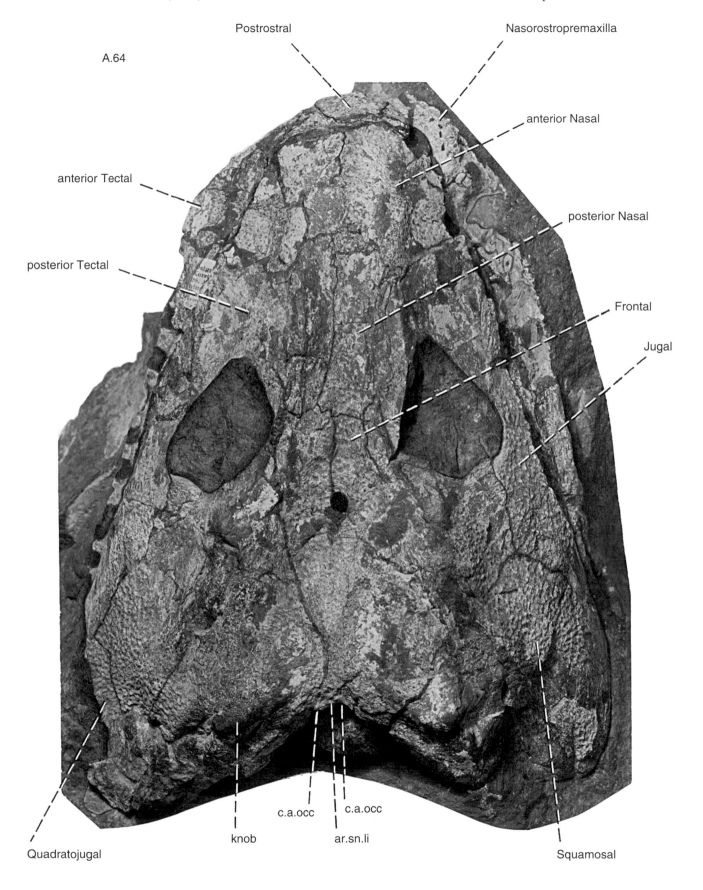

A.64

Postrostral

Nasorostropremaxilla

anterior Nasal

anterior Tectal

posterior Nasal

posterior Tectal

Frontal

Jugal

c.a.occ

c.a.occ

knob

ar.sn.li

Quadratojugal

Squamosal

Plate 4

□1, 2. Palate and counterpart of A. 64 $\times^2/_3$.

□3. Median tubular portion of otoccipital in ventral view. $\times^4/_3$.

□4. Skull in lateral view. All same specimen as in Pls. 2:2 and 3. $\times^2/_3$.

bcr.mu, posterior area for attachment of basicranial muscles; *c.palt*, canals for terminal twigs of r.palatinus VII; *c.v.ju*, canal for jugular vein; *f.ac*, foramen for n.acousticus; *fe.exch*, fenestra exochoanalis; *fo.mu.add*, fossa for adductor muscle; *gr.fi.occ, gr.fi.ot.va, gr.fi.ot.vp*, grooves probably marking position of occipital, and anterior and posterior ventral otical fissurae; *pr.asc*, processus ascendens of parasphenoid.

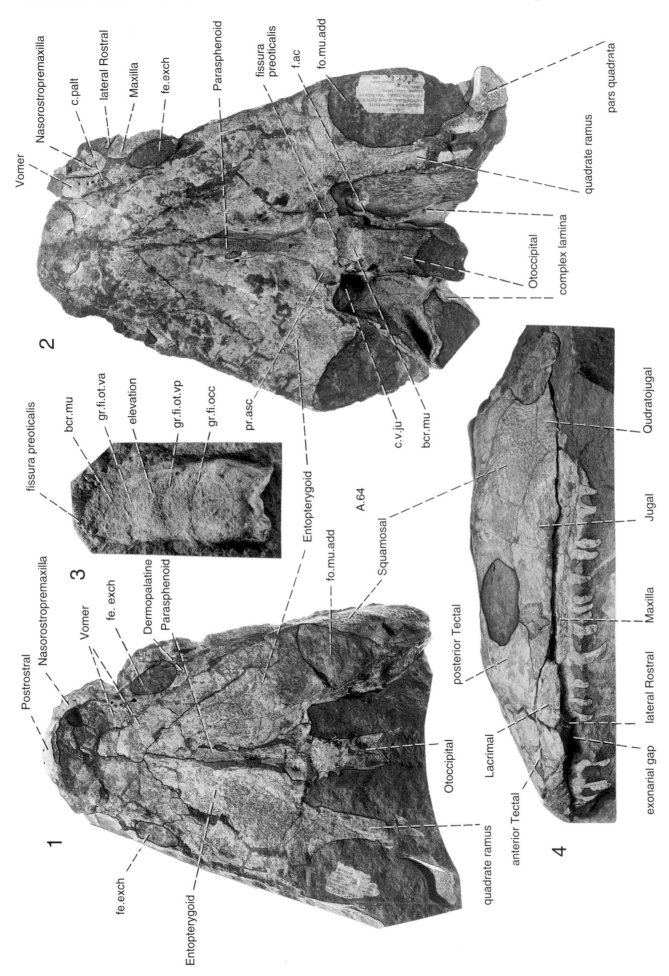

Plate 5

□1–4. Imperfect skull tables of four specimens (A. 193–195, 199), coll. 1951, in horizon at 980 m in the Lower Red Division of the Remigolepis Series (Group), Smith Woodward Bjerg (A. 194 also in Fig. 22D). Natural size.

ioc, groove for infraorbital sensory canal.

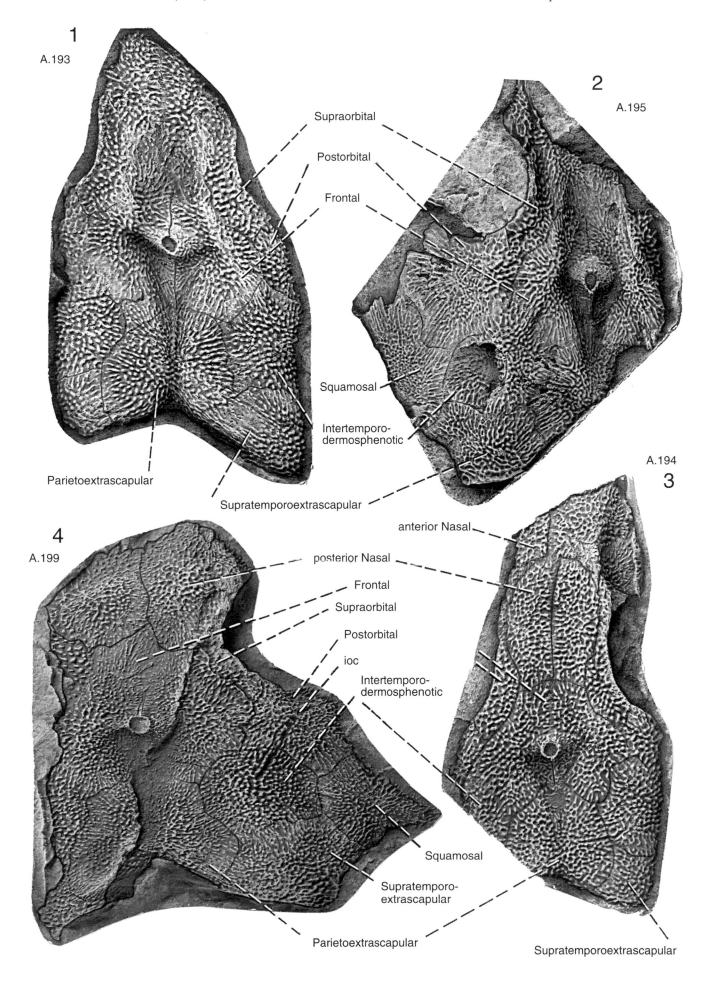

1 A.193

Supraorbital

Postorbital

Frontal

2 A.195

Squamosal

Intertemporo-
dermosphenotic

Parietoextrascapular

Supratemporoextrascapular

A.194

3

anterior Nasal

4 A.199

posterior Nasal

Frontal

Supraorbital

Postorbital

ioc

Intertemporo-
dermosphenotic

Squamosal

Supratemporo-
extrascapular

Parietoextrascapular

Supratemporoextrascapular

Plate 6

☐1. Latex cast of imperfect skull in dorsal view. A. 28, coll. 1932, Celsius Bjerg. Natural size.

☐2, 3, Anterior and posterior parts of skull table of A. 55 (same specimen as in Pls. 13:1, 26:1, 33:2), coll. 1934, Sederholm Bjerg (1174 m). Natural size.

ioc, groove for infraorbital sensory canal; *stcc*, supratemporal commissural sensory canal (prepared by Säve-Söderbergh).

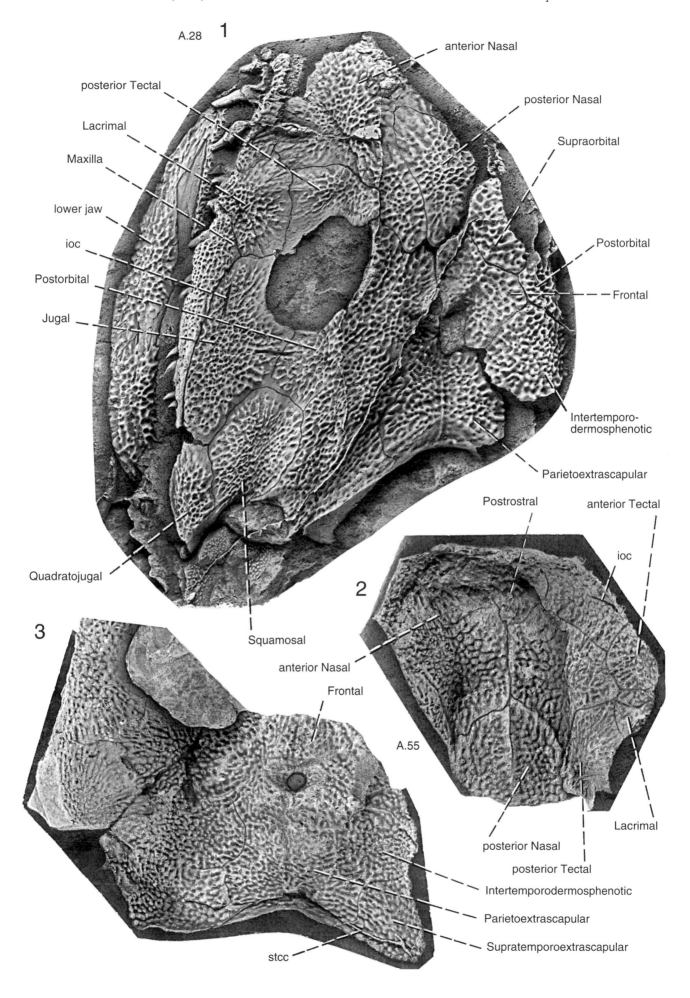

A.28

1

anterior Nasal

posterior Tectal

Lacrimal

Maxilla

lower jaw

ioc

Postorbital

Jugal

posterior Nasal

Supraorbital

Postorbital

Frontal

Intertemporo-
dermosphenotic

Parietoextrascapular

Quadratojugal

Squamosal

2

Postrostral

anterior Tectal

ioc

A.55

anterior Nasal

Frontal

Lacrimal

posterior Nasal

posterior Tectal

Intertemporodermosphenotic

Parietoextrascapular

Supratemporoextrascapular

3

stcc

Plate 7

☐1, 2. Imperfect narrow palate and part of skull table in dorsal and lateral views. A. 102, coll. 1948, Celsius Bjerg. Natural size.

fe.exch, fenestra exochoanalis; *fo.mu.add*, fossa for adductor muscle; *ioc, p-mc*, infraorbital and preoperculo-mandibular sensory canals.

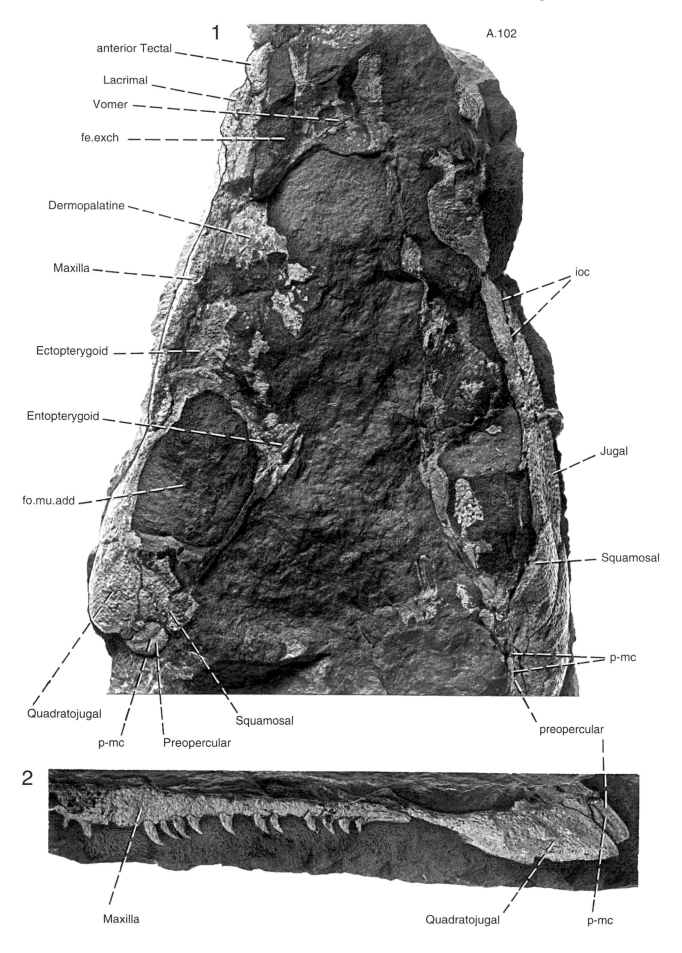

1

A.102

anterior Tectal
Lacrimal
Vomer
fe.exch

Dermopalatine

Maxilla

Ectopterygoid

Entopterygoid

fo.mu.add

ioc

Jugal

Squamosal

p-mc

Quadratojugal

p-mc　Preopercular

Squamosal

preopercular

2

Maxilla

Quadratojugal　p-mc

Plate 8

☐1, 2. Part and counterpart of skull in dorsal view. A. 148, coll. 1949, Smith Woodward Bjerg.

☐3. Latex cast of part of skull table. A. 57, coll. 1934, Sederholm Bjerg (1174 m). Natural size.

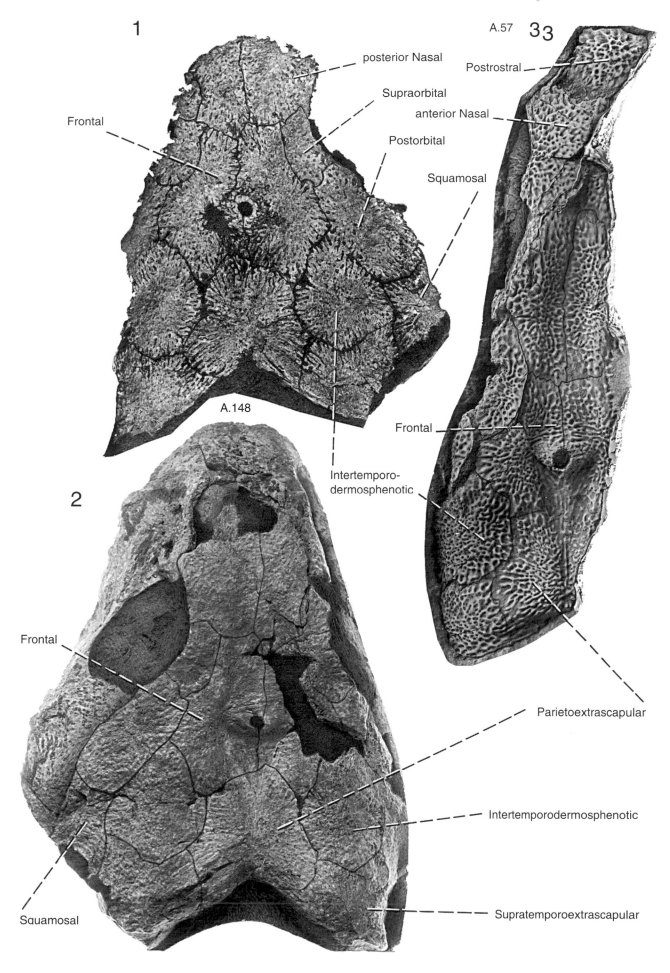

1

posterior Nasal

Supraorbital

Frontal

Postorbital

Squamosal

A.148

Intertemporo-
dermosphenotic

2

Frontal

Frontal

Squamosal

A.57 33

Postrostral

anterior Nasal

Frontal

Parietoextrascapular

Intertemporodermosphenotic

Supratemporoextrascapular

Plate 9

☐1, 2. Parts of skull table in lateral view. A. 57 (same specimen as in Pl. 8:3). Natural size.

☐3. Imperfect skull table with sensory canals prepared by Säve-Söderbergh. A. 63, coll. 1934, Sederholm Bjerg (1174 m). Natural size.

eth.com, ethmoidal commissure: *ioc, juc, p-mc*, infraorbital, jugal and preoperculo-mandibular canals.

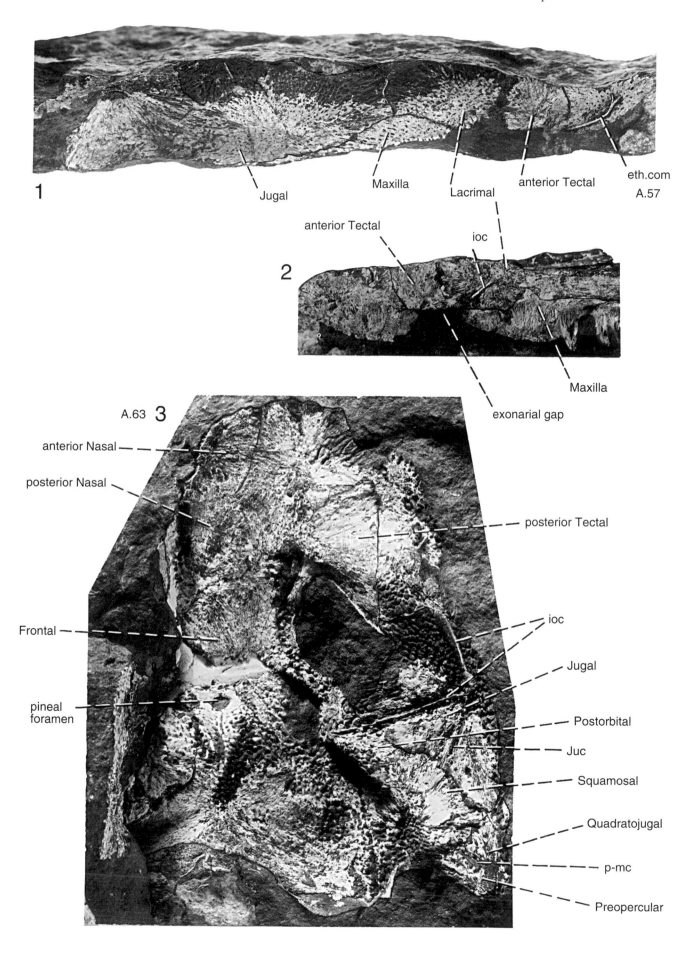

1

Jugal

Maxilla

Lacrimal

anterior Tectal

eth.com

A.57

2

anterior Tectal

ioc

Maxilla

exonarial gap

A.63 3

anterior Nasal

posterior Nasal

posterior Tectal

Frontal

ioc

Jugal

pineal foramen

Postorbital

Juc

Squamosal

Quadratojugal

p-mc

Preopercular

Plate 10

Part of skull showing the lateral rostral. A.64 (same specimen as in Pls. 3, 4). ×6.

☐1. Part of palate (cf. Pl. 4:2)

☐2. Part of skull in lateral view.

☐3. Counterpart of the part with the lateral rostral shown in 2.

c.palt, canals for twigs of r.palatinus VII.

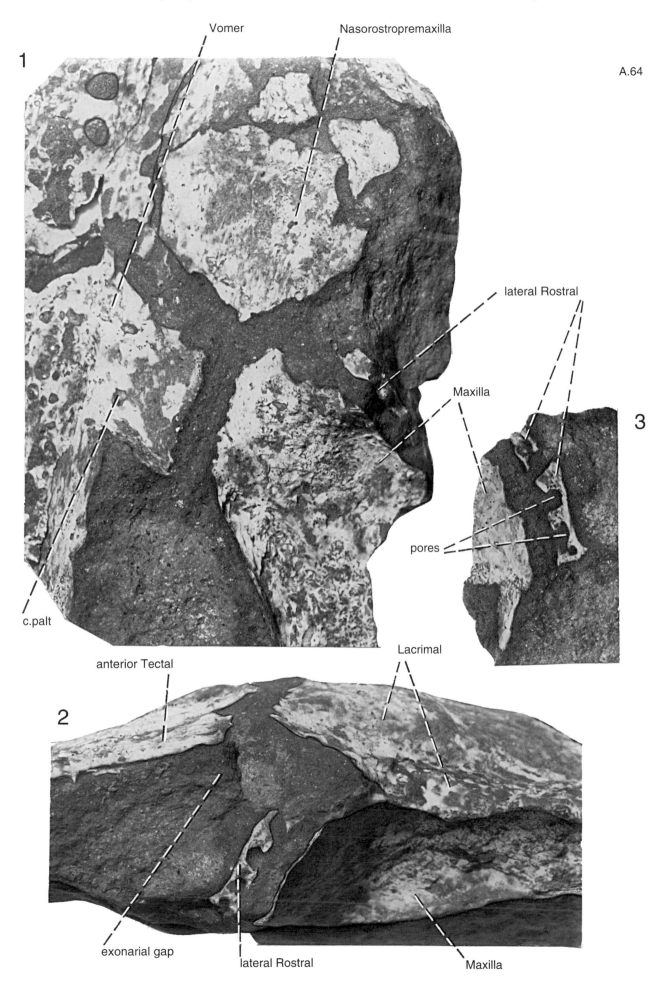

A.64

Plate 11

☐1, 2. Anterior part of snout in ventral and anterior views. A. 94, coll. 1948, Celsius Bjerg. ×2.

☐3. Part of skull table. A. 240, coll. 1955, Smith Woodward Bjerg. Natural size.

☐4. Latex cast of anterior part of skull. A. 86, coll. 1947, Celsius Bjerg. Natural size.

a.pal.fe, anterior palatal fenestra; *c.palt*, canals for twigs of r.palatinus VII; *fe.exch*, fenestra exochoanalis; *for*, foramen.

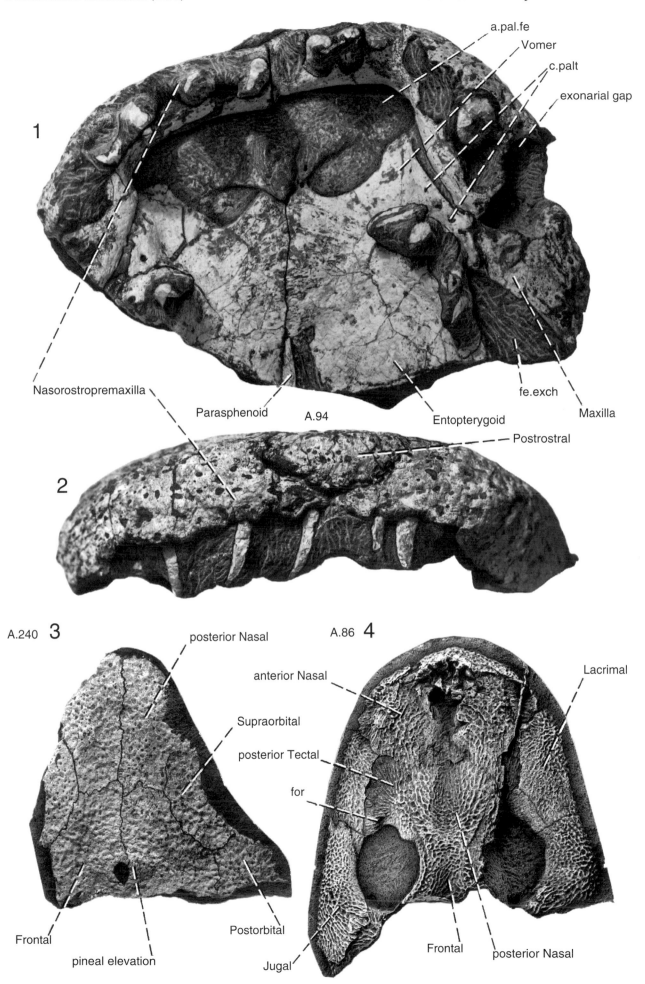

1

a.pal.fe
Vomer
c.palt
exonarial gap

Nasorostropremaxilla

Parasphenoid   A.94

fe.exch

Maxilla

Entopterygoid

2

Postrostral

A.240 3

posterior Nasal

anterior Nasal

Lacrimal

Supraorbital

posterior Tectal

for

Frontal

pineal elevation

Postorbital

Jugal

Frontal

posterior Nasal

A.86 4

Plate 12

Skull laterally compressed in (1) lateral, (2) posterior and (3) ventral views. A. 251, coll.1955, Celsius Bjerg. Natural size. In 1, note vertical row of three pores in quadratojugal (see p. 100).

c.a.occ, canals for occipital arteries; *vert.la*, complex vertical lamina (exoskeletal part).

A.251 **1**

pineal foramen

Supraorbital

orbital fenestra

posterior Tectal

anterior Nasal

c.a.occ

vert.la

broad ridge

Preopercular

Quadratojugal

Jugal

lower jaw

Nasorostropremaxilla

2

Parietoextrascapular

pineal foramen

Supratemporo-extrascapular

c.a.occ

broad ridge

vert.la

3

lower jaw

Plate 13

☐1. Posteroventral part of skull table in lateral view. A. 55, same specimen in Pls. 6:2, 3; 26:1; 33:2.

☐2. Posterior part of skull in lateral view. A. 113, coll. 1948, Stensiö Bjerg. Natural size.

☐3–5. Posterior part of skull with subopercular of (4) right and (3, 5) left sides. A. 89, coll. 1948, Stensiö Bjerg. Natural size.

ioc, infraorbital sensory canal.

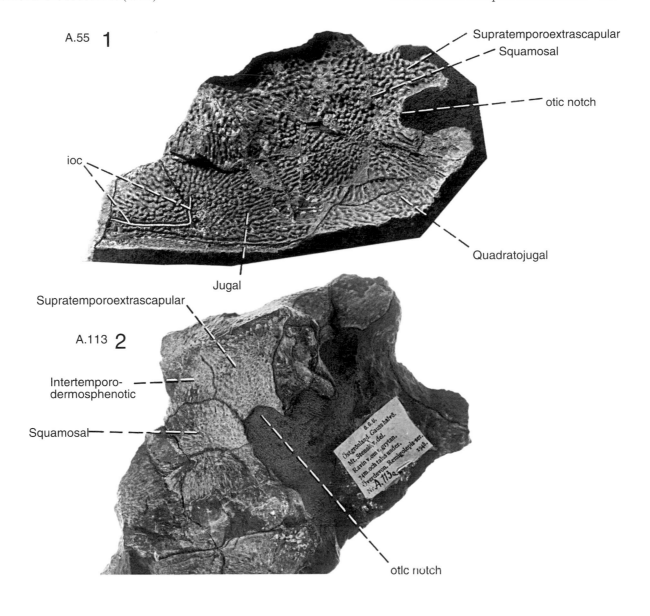

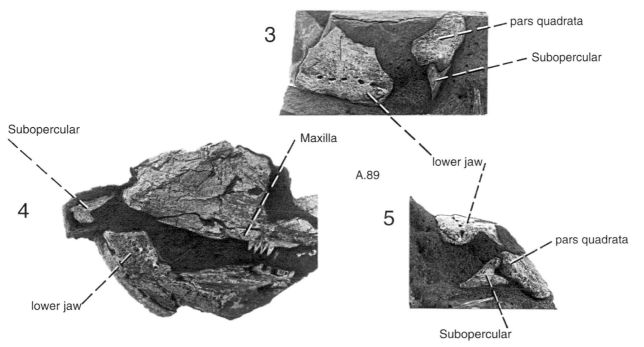

Plate 14

Skull table in ventral view. A. 158, coll. 1949, Sederholm
Bjerg (1174 m). Natural size.

A.158

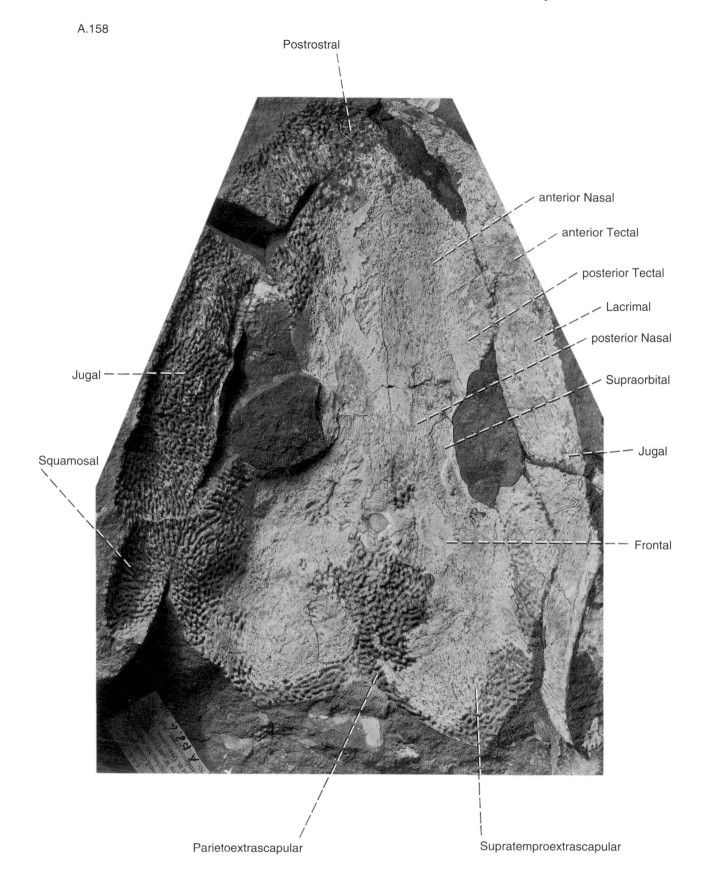

Postrostral

anterior Nasal

anterior Tectal

posterior Tectal

Lacrimal

posterior Nasal

Supraorbital

Jugal

Jugal

Squamosal

Frontal

Parietoextrascapular

Supratemproextrascapular

Plate 15

Skull in dorsal view after removal of main part of skull table.
Same specimen as in Pls. 14, 16–18. $\times \frac{3}{2}$.

sp.pr.S.-S, space for the process of Säve-Söderbergh.

A.158

Nasorostropremaxilla

Ethmospenoid

anterior Tectal

posterior Tectal

Lacrimal

parotic groove

parotic ridge

sp.pr.S.-S

Otoccipital

otic elevations

sp.pr.S.-S

Plate 16

Posterior division of skull shown in Pl. 15. $\times^3/_2$.

☐1. Ventral view.

☐2. Right half in ventromedial view.

☐3, 4. Longitudinal sections.

☐5. Transverse section through left sacculus vesicle.

bcr.mu, posterior attachment area for basicranial muscles; *pr.Fr*, ventral process of frontal; *pr.S.-S*, Säve-Söderbergh's ventral process; *sp.pr.S.-S*, space for this process; *VIII*, canal for n.acousticus.

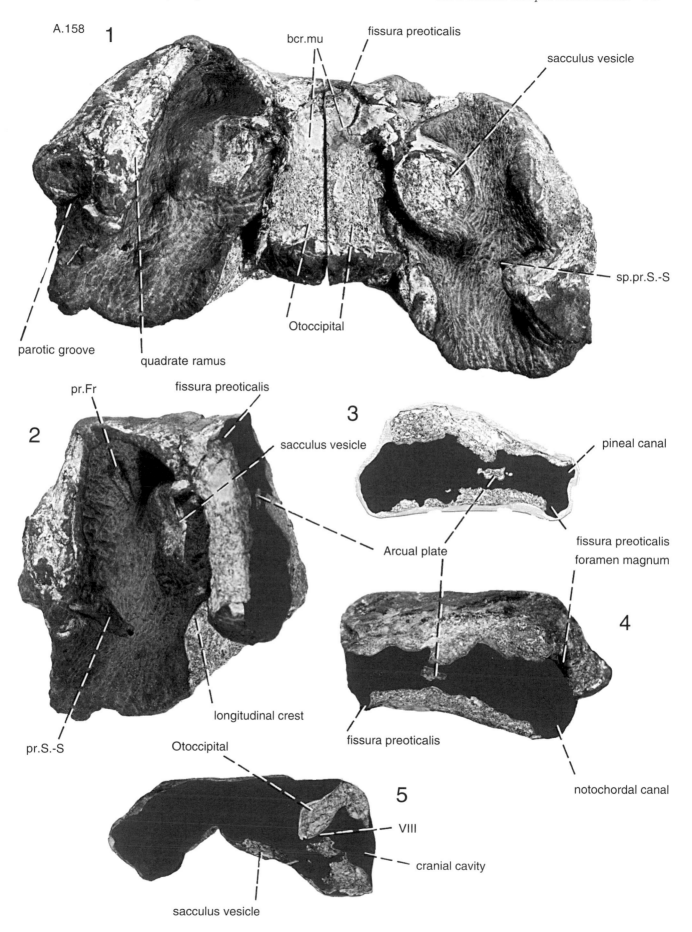

A.158

1

bcr.mu

fissura preoticalis

sacculus vesicle

sp.pr.S.-S

Otoccipital

parotic groove

quadrate ramus

2

pr.Fr

fissura preoticalis

sacculus vesicle

Arcual plate

longitudinal crest

pr.S.-S

Otoccipital

sacculus vesicle

3

pineal canal

fissura preoticalis
foramen magnum

4

fissura preoticalis

notochordal canal

5

VIII

cranial cavity

Plate 17

□1–3. Same specimen as in Pls. 14–16, 18. Natural size.

□1. Palate in dorsal view.

□2. Left side in ventral view.

□3. Skull in lateral view.

NRP, Nasorostropremaxilla; *c.palt*, canals for twigs of r.pala-
tinus VII.

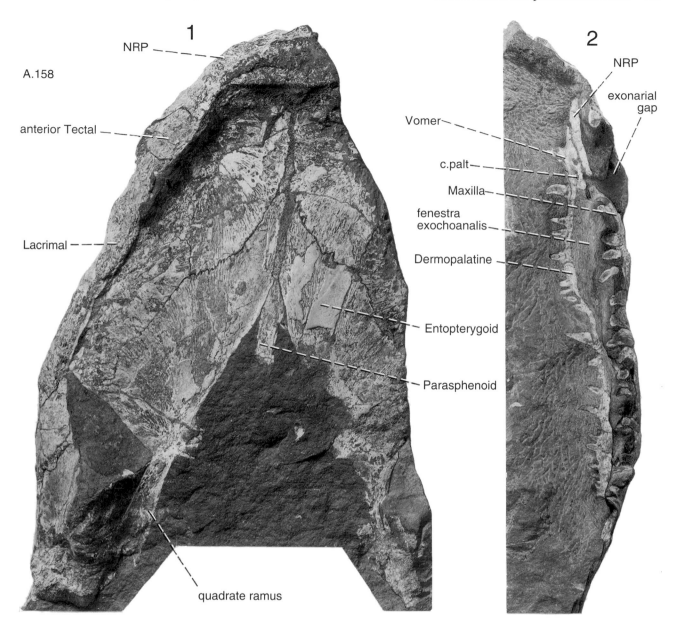

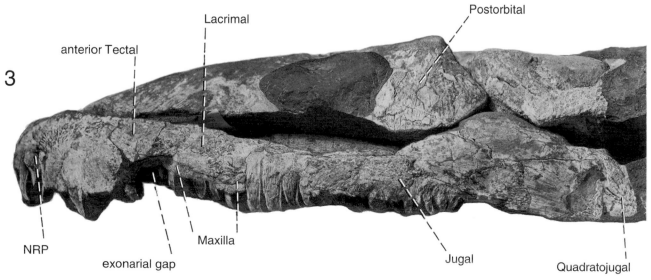

Plate 18

Counterpart of part of the palate in Pl. 17:1, with grooves for nerves and blood vessels on the dorsal side of the periosteal lining of the palatoquadrate. Ventral view. ×2.

c.v.ju, canals for jugular veins; *end.pq*, endoskeleton of palatoquadrate; *f.bh*, buccohypophysial foramen; *gr.r.md, gr.r.mx*, grooves for r.mandibularis and r.maxillaris V; *per.li*, periosteal lining.

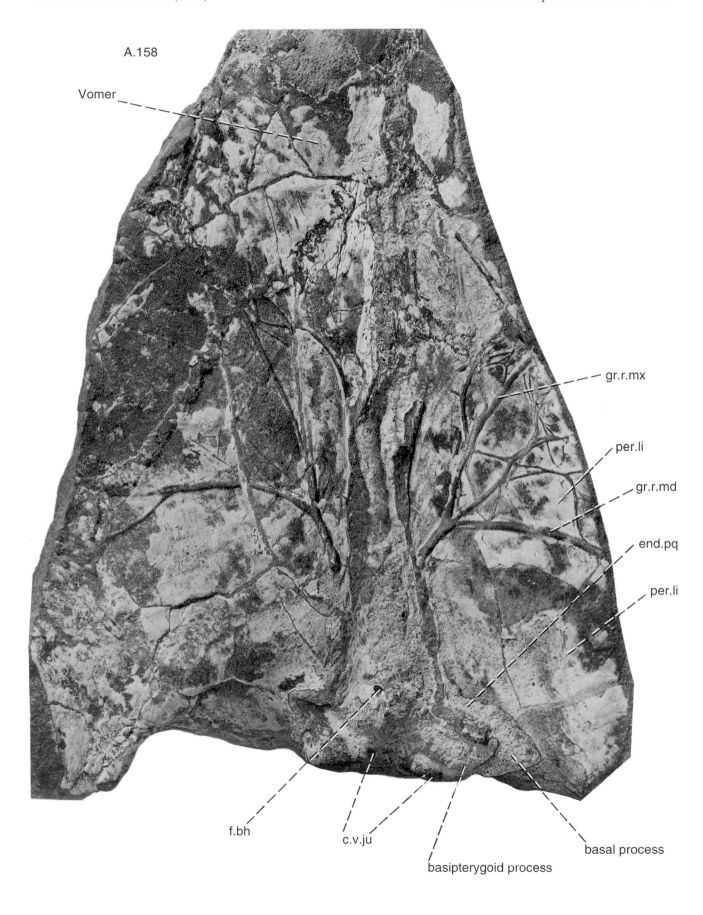

A.158

Vomer

gr.r.mx

per.li

gr.r.md

end.pq

per.li

f.bh

c.v.ju

basipterygoid process

basal process

Plate 19

A. 60, coll. 1934, Sederholm Bjerg (1174 m).

☐1. Otoccipital in ventral view. ×2.

☐2. Left posterolateral part of skull in posterior view. Natural size.

☐3. Right half of skull with part of left entopterygoid in lateral view. Natural size.

☐4. Right side of otoccipital in lateral view. ×2.

can, narrow canal; *for*, foramina; *VIII*, foramen for. n. acousticus.

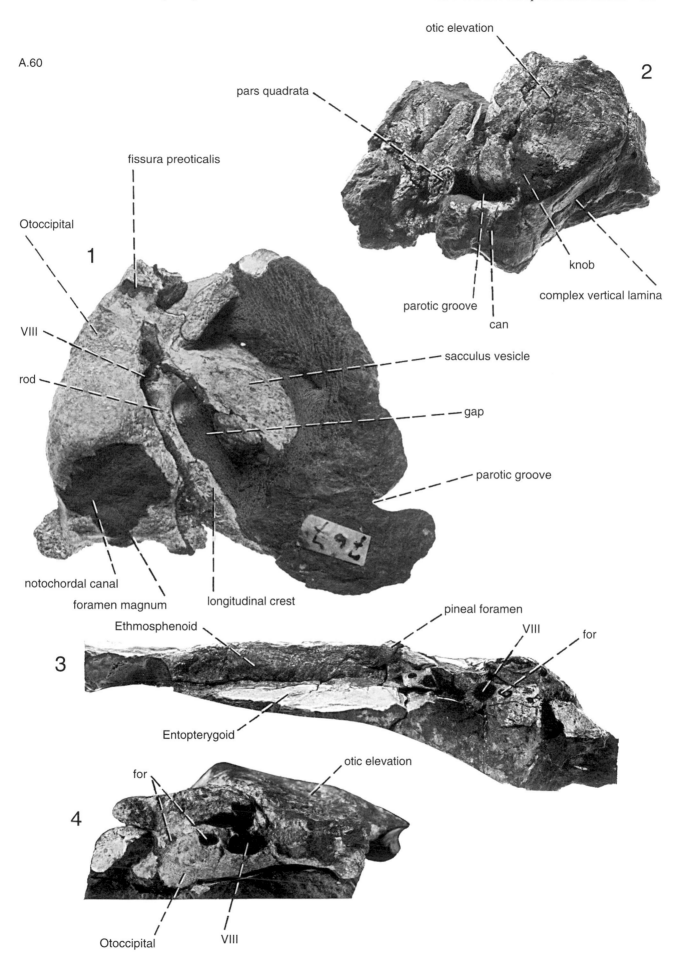

A.60

Plate 20

□1, 2. Posterior parts of palate in ventral views. A. 53, A. 63, coll. 1934, Sederholm Bjerg (1174 m). Natural size.

□3, 4. Right side of otoccipital with counterpart. A. 139, coll. 1949, Sederholm Bjerg (1174 m). Same specimen as in Pl. 24. ×2.

c.v.ju, canal for jugular vein; *f.bh*, buccohypophysial foramen; *fo.mu.add*, fossa for adductor muscle; *for*, foramina; *gr.pot*, parotic groove; *pr.bas*, processus basalis; *pr. S.-S*, the process of Säve-Söderbergh; *sp.pr.S.-S*, space for that process.

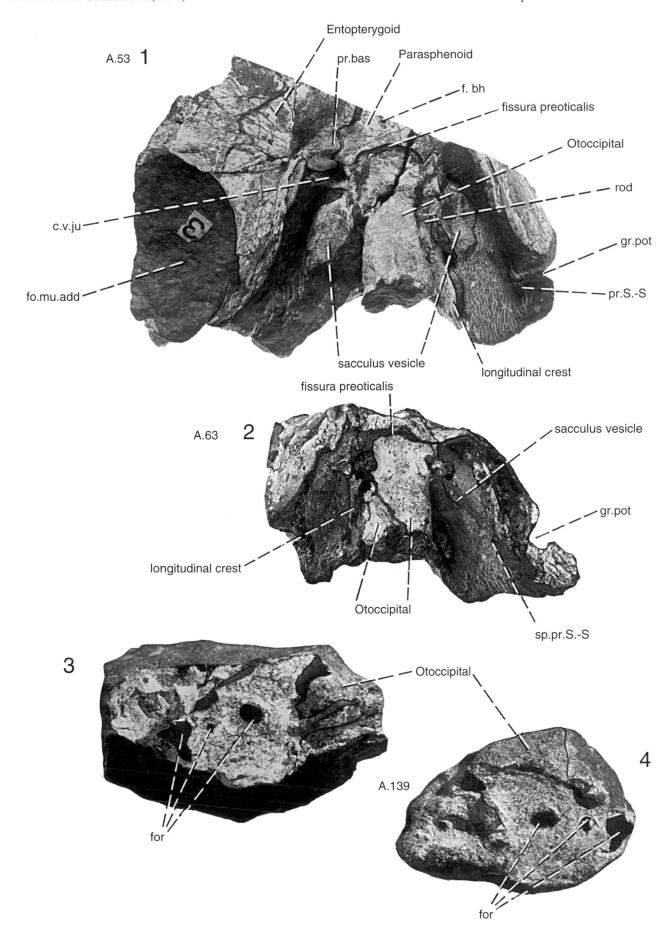

Plate 21

☐1–3. Longitudinal section through otoccipital (1, 2) and part of palate in ventral view (3). A. 62, coll. 1934, Sederholm Bjerg (1174 m). Natural size.

☐4. Part of palate in ventral view. A. 72, coll. 1936, Sederholm Bjerg. Natural size.

☐5. Posterior part of skull in lateral view. A. 115, coll. 1948, Smith Woodward Bjerg. ×³⁄₂. Same specimen as in Pls. 42, 45, 46, 53, 54.

bcr.mu, anterior and posterior areas of attachment for basicranial muscle; *f.bh*, buccohypophysial foramen; *hor.pl*, horizontal plate of paratemporal process; *ipter.vac*, interpterygoid vacuity; *i.V*, notch for n.trigeminus; *pr.asc.P*, *pr.asc.pq*, processus ascendens of parasphenoid and palatoquadrate; *pr.pt*, paratemporal process; *str.pr*, striated processes; *vert.la*, complex vertical lamina.

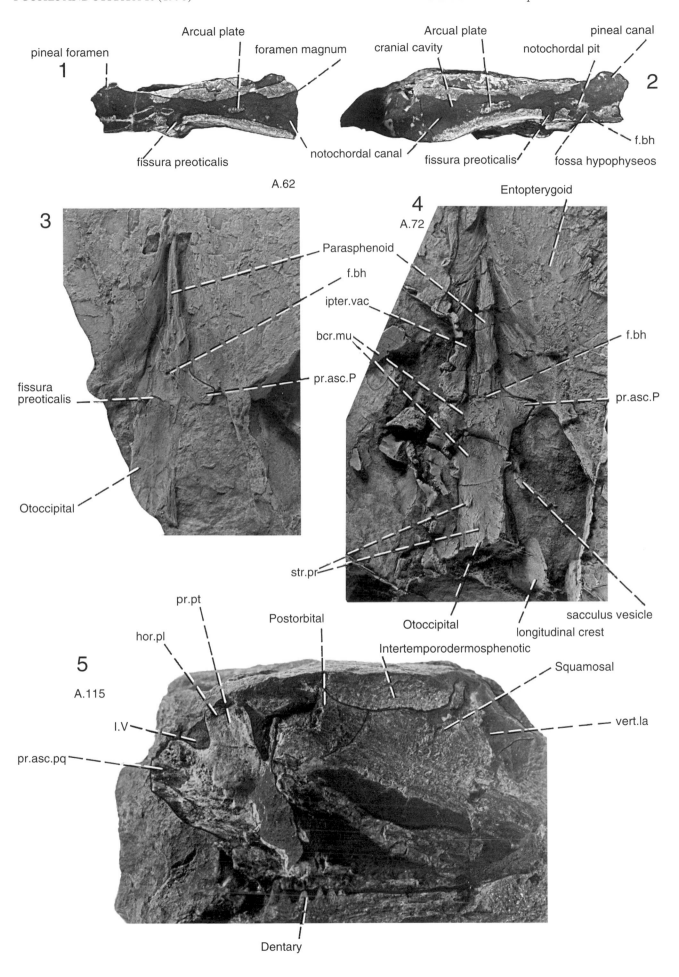

1 — pineal foramen, Arcual plate, foramen magnum, fissura preoticalis, notochordal canal. A.62

2 — cranial cavity, Arcual plate, pineal canal, notochordal pit, f.bh, fossa hypophyseos, fissura preoticalis

3 — Parasphenoid, f.bh, pr.asc.P, fissura preoticalis, Otoccipital

4 — Entopterygoid, ipter.vac, bcr.mu, f.bh, pr.asc.P, str.pr, Otoccipital, longitudinal crest, sacculus vesicle. A.72

5 — pr.pt, hor.pl, Postorbital, Intertemporodermosphenotic, Squamosal, I.V, pr.asc.pq, vert.la, Dentary. A.115

Plate 22

A. 5 (*Ichthyostegopsis wimani*, Säve-Söderbergh 1932a, Pls. 16:1, 17–19), coll. 1931, Celsius Bjerg.

☐1. Skull in posterior view. ×2.

☐2. Part of palatoquadrate in anterolateral view. ×3.

☐3. Incomplete palate in ventral view. ×3/2.

can, narrow canal; *c.a.occ*, canals for occipital arteries; *fe.exch*, fenestra exochoanalis; *f.bh*, buccohypophysial foramen; *fo.mu.add*, fossa for adductor muscle; *hor.pl*, horizontal plate of paratemporal process; *i.v.io*, notch for infraorbital vein; *i.V*, notch for n.trigeminus; *pr.asc*, processus ascendens; *pr.pt*, paratemporal process; *pr.S.-S*, Säve-Söderbergh's ventral process; *tr.w*, transverse wall; *vert.la*, complex vertical lamina.

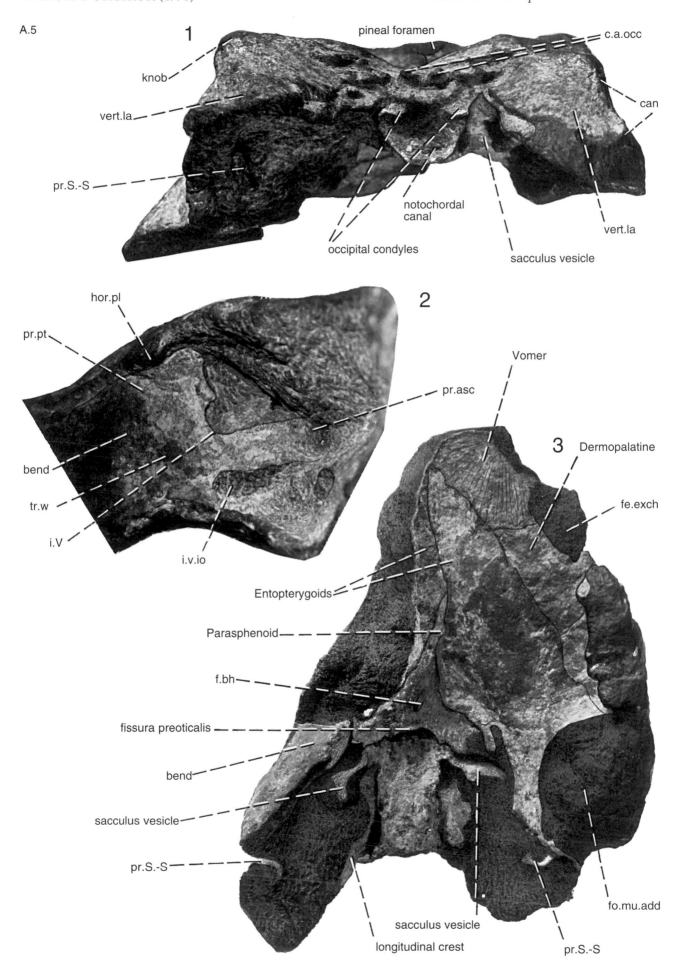

A.5

Plate 23

Imperfect skull. A. 152, coll. 1949, Sederholm Bjerg (1174 m). Same specimen as A. 59, coll. 1934.

☐1. Skull in dorsal view. Natural size.

☐2. Posterior part of palate. Natural size.

☐3, 4. Posterior part of skull in lateral view. Natural size. In 4 the connecting process shown in 3 has been removed.

☐5. Latex cast of the parotic ridge and the space for Säve-Söderbergh's ventral process. ×2.

can, narrow canal; *conn.pr*, connecting process; *fe.exch*, fenestra exochoanalis; *long.cr*, longitudinal crest; *pr.bas*, processus basalis; *pr.S.-S*, Säve-Söderbergh's ventral process; *sp.long.cr*, space for dorsal end of longitudinal crest; *sp.pr.Fr*, space for ventral process of frontal; *sp.pr.S.-S*, space for Säve-Söderbergh's ventral process; *vert.la*, complex vertical lamina.

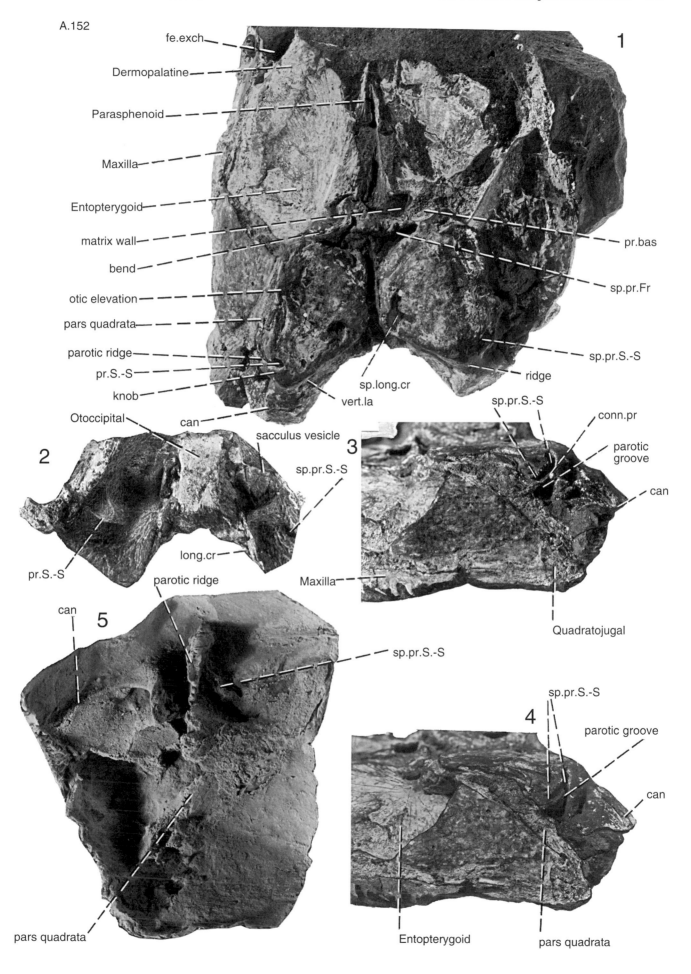

A.152

Plate 24

Imperfect skull, A. 139, coll. 1949, Sederholm Bjerg (1174 m). Natural size. Same specimen as in Pl. 20:3, 4.

☐1, 2. Left side of skull with latex cast in lateral view.

☐3. Latex cast of right side.

☐4. Part of palate.

fo.mu.add, fossa for adductor muscle; *f.r.aut.V*, foramen for r.auriculotemporalis V; *gr.v.ju*, groove for jugular vein; *str.pr*, striated processes.

A.139

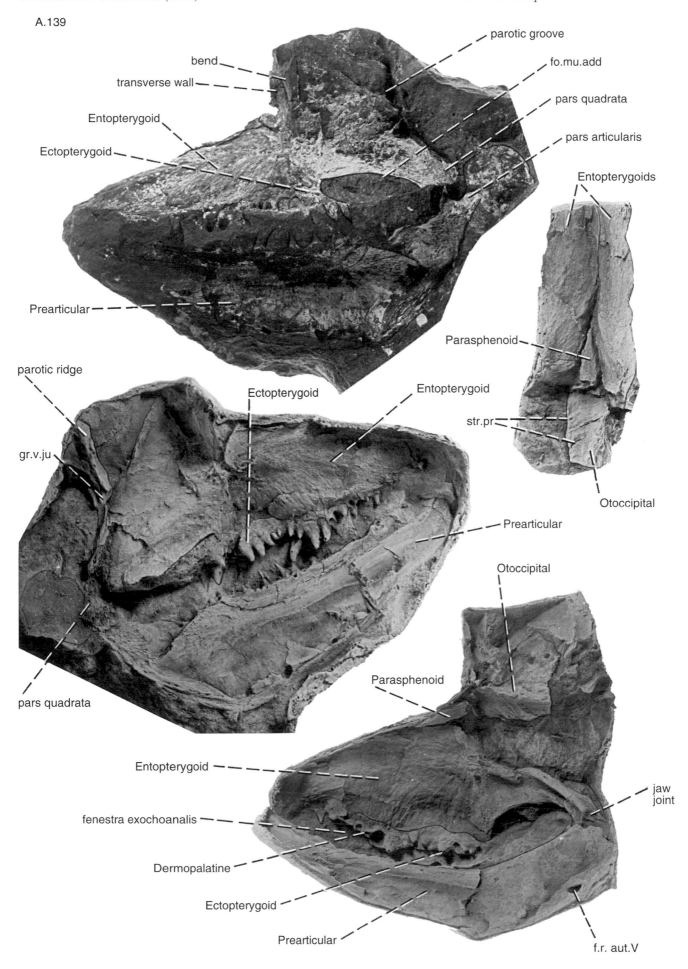

parotic groove

bend

fo.mu.add

transverse wall

pars quadrata

Entopterygoid

pars articularis

Ectopterygoid

Entopterygoids

Prearticular

Parasphenoid

parotic ridge

Ectopterygoid

Entopterygoid

gr.v.ju

str.pr

Prearticular

Otoccipital

pars quadrata

Otoccipital

Parasphenoid

Entopterygoid

jaw
joint

fenestra exochoanalis

Dermopalatine

Ectopterygoid

Prearticular

f.r. aut.V

Plate 25

□1–3. Posterior part of skull in dorsal, ventral and posterior views. A. 117, coll. 1949, Smith Woodward Bjerg. 1, Natural size. 2, 3 ×$\frac{3}{2}$.

ar.sn.li, area for attachment of supraneural ligament; *c.a.occ*, canal for occipital artery; *f.bh*, buccohypophysial foramen; *fl.Enpt*, flange of quadrate ramus of entopterygoid; *i.X*, notch for n.vagus; *la.Sq*, squamosal lamina; *pr.bas*, processus basalis; *pr.dl*, dorsolateral process of otoccipital; *sp.pr.Fr*, space for ventral process of frontal; *pr.S.-S.la*, *pr.S.-S.r*, laminar and rodlike portions of Säve-Söderbergh's ventral process; *vert.la*, vertical complex lamina; *VIII*, foramen for n.acousticus.

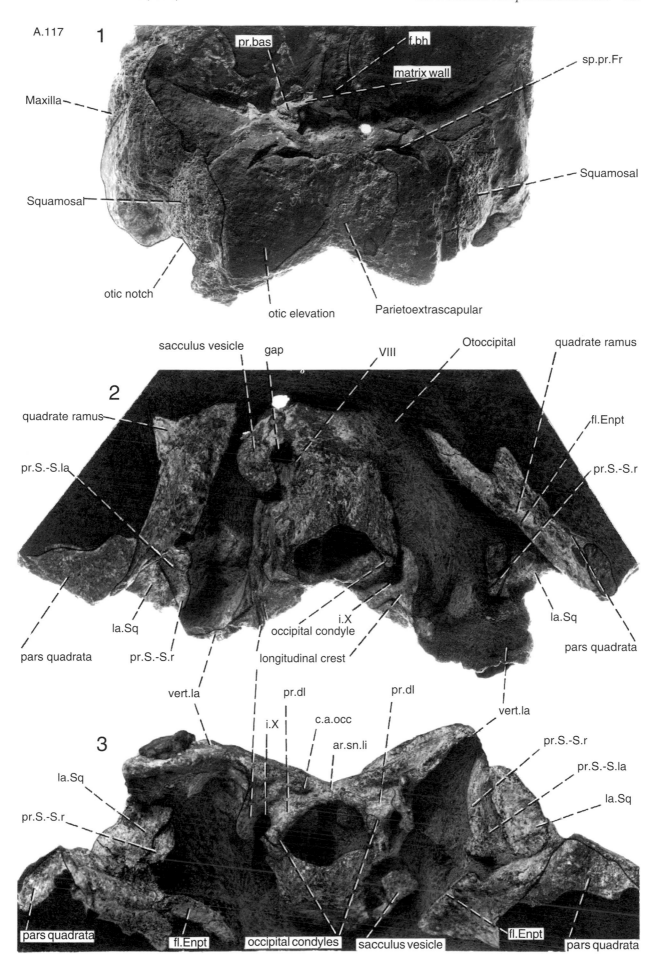

Plate 26

□1. Palate in ventral view. A. 55, same specimen as in Pls. 6:2, 3; 13:1; 33:2. Natural size.

□2. Posterior part of palate. A. 53, coll. 1934, Sederholm Bjerg (1174 m). Natural size.

NRP, nasorostropremaxilla; *art.cond*, articular condyles; *bcr.mu*, anterior and posterior areas of attachment for basicranial muscles; *c.palt*, canals for twigs of r.palatinus; *f.bh*, buccohypophysial foramen; *fe.exch*, fenestra exochoanalis; *fi.prot*, fissura preoticalis; *fl.Enpt.* flange of quadrate ramus of entopterygoid; *fo.mu.add*, fossa for adductor muscle; *la.Sq*, squamosal lamina; *pons.nch*, pons nariochoanalis; *pr.asc*, processus ascendens of parasphenoid; *pr.bas*, processus basalis; *pr.dl*, dorsolateral process of otoccipital; *pr.S.-S.la*, *pr.S.-S.r*, laminar and rodlike portions of the process of Säve-Söderbergh; *str.pr*, striated processes.

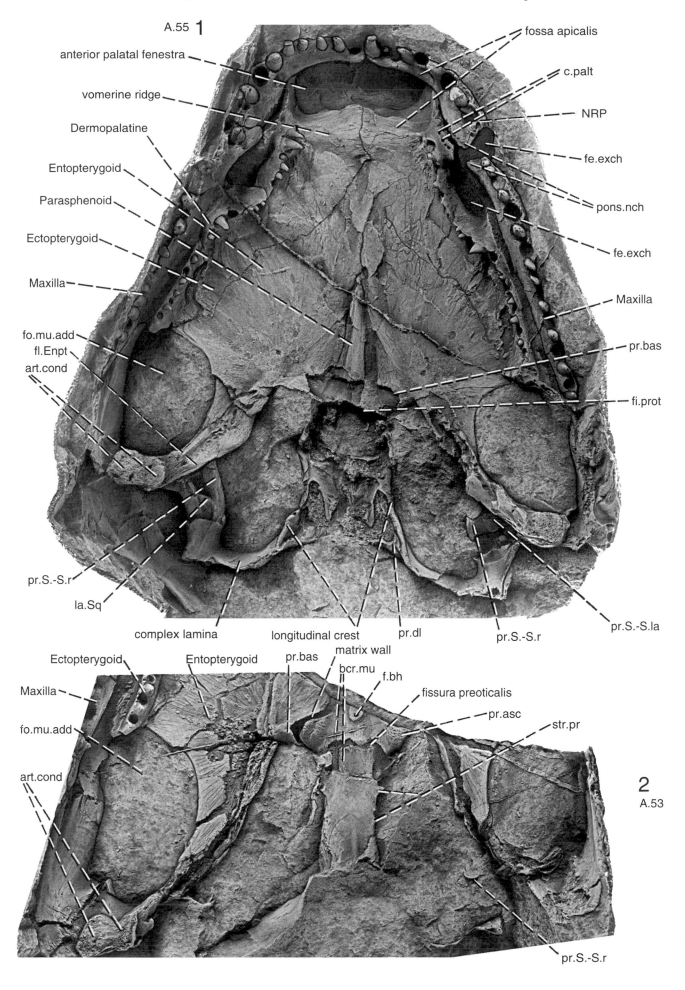

A.55 **1**

anterior palatal fenestra

vomerine ridge

Dermopalatine

Entopterygoid

Parasphenoid

Ectopterygoid

Maxilla

fo.mu.add
fl.Enpt
art.cond

pr.S.-S.r

la.Sq

fossa apicalis

c.palt

NRP

fe.exch

pons.nch

fe.exch

Maxilla

pr.bas

fi.prot

complex lamina

longitudinal crest
matrix wall

pr.dl

pr.S.-S.r

pr.S.-S.la

Ectopterygoid

Maxilla

fo.mu.add

art.cond

Entopterygoid

pr.bas

bcr.mu

f.bh

fissura preoticalis

pr.asc

str.pr

2

A.53

pr.S.-S.r

Plate 27

☐1. Latex cast of broad palate. A. 70, coll. 1936, Sederholm
Bjerg. Natural size.

☐2. Latex cast of posterior part of palate. A. 58, coll. 1934,
Sederholm Bjerg (1174 m). Natural size.

bcr.mu, anterior and posterior areas of attachment for basicranial muscles; *f.bh*, buccohypophysial foramen; *fo.mu.add*,
fossa for adductor muscle; *ipter.vac*, interpterygoid vacuity;
pr.asc, processus ascendens; *ri.P*, ridge of parasphenoid;
str.pr, striated processes.

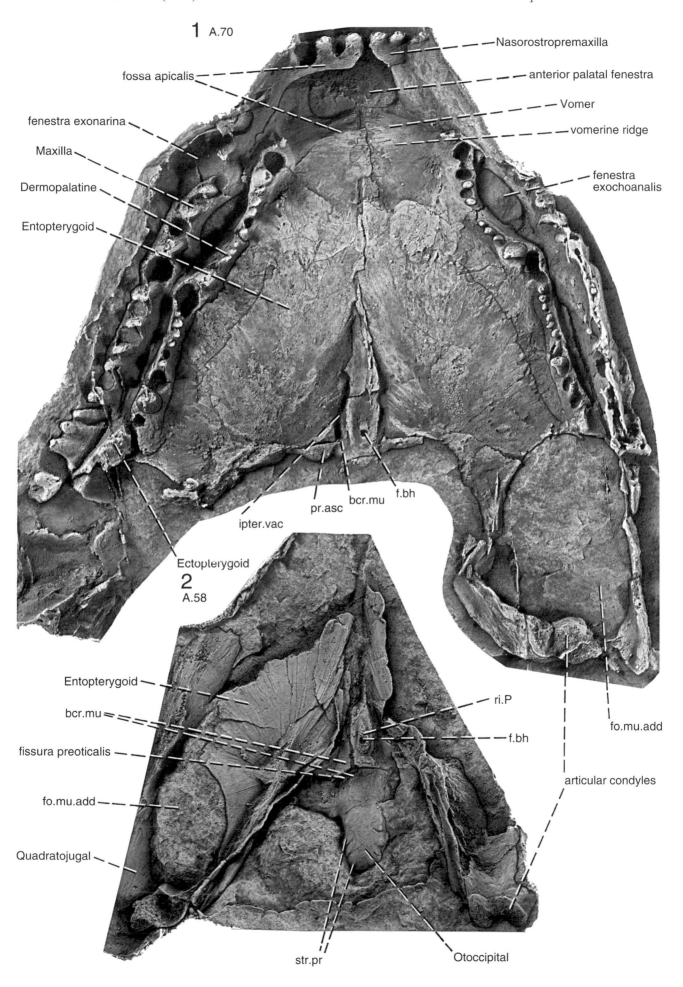

1 A.70

Nasorostropremaxilla

fossa apicalis

anterior palatal fenestra

Vomer

fenestra exonarina

vomerine ridge

Maxilla

fenestra exochoanalis

Dermopalatine

Entopterygoid

ipter.vac

pr.asc

bcr.mu

f.bh

Ectopterygoid

2 A.58

Entopterygoid

ri.P

bcr.mu

f.bh

fissura preoticalis

fo.mu.add

articular condyles

Quadratojugal

fo.mu.add

str.pr

Otoccipital

Plate 28

Latex cast of part of palate. A. 234, coll. 1955, Sederholm Bjerg (1174 m). ×3.

bcr.mu, anterior and posterior areas of attachment for basicranial muscles; *f.bh*, buccohypophysial foramen; *fi.prot*, fissura preoticalis; *fo.mu.add*, fossa for adductor mucle; *pr.asc*, processus ascendens; *pr.bas*, processus basalis; *ri.P*, ridge of parasphenoid; *str.pr*, striated processes.

A.234

Entopterygoid

ri.P

pr.asc

bcr.mu

pr.bas

fi.prot

fo.mu.add

quadrate ramus

str.pr

Otoccipital

Plate 29

Latex cast of disarticulated palatal bones. A. 149, coll. 1949, Sederholm Bjerg (1174 m). Natural size.

stri.Enpt, striated areas of entopterygoid bordering interpterygoid vacuity.

A.149

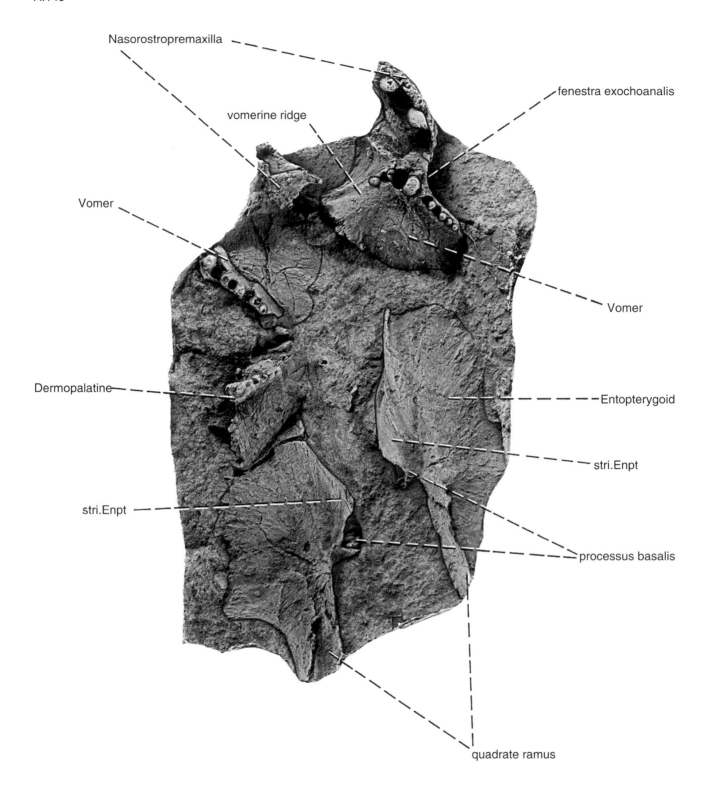

Plate 30

☐1–3. Incomplete lower jaw in lateral, medial and dorsal views. A. 84, coll. 1936, Sederholm Bjerg. ×³⁄₂.

Co.2, Co.3, coronoids 2 and 3; *Id.2, 3, 4*, infradentaries 2–4; *Mb.art*, articular pits of meckelian bone; *Prart*, prearticular; *f.r.aut.V*, foramen for r.auriculotemporalis V; *f.r.md.VII*, foramen for r.mandibularis of n.facialis (chorda tympani); *od.De*, area overlapped by dentary; *orc.p*, pore of oral sensory canal; *p-mc*, preoperculo-mandibular canal; *p-mc.p*, pores of that canal; *pr.retr*, retroarticular process.

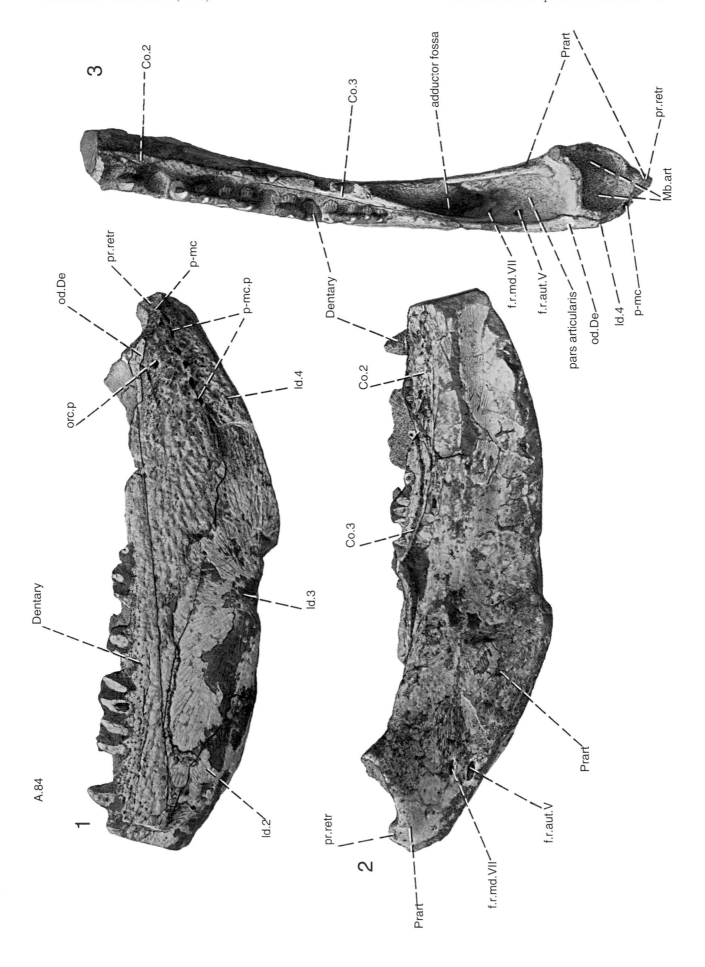

Plate 31

☐1, 2. Lower jaws in dorsal view and right lower jaw in medial view. A. 138, coll. 1949, Sederholm Bjerg (1174 m). Natural size.

☐3. Lower jaw in medial view. A. 57, coll. 1934, Sederholm Bjerg (1174 m). Natural size.

Co.1, 2, 3, coronoids 1–3; *Id.1, Id.3,* infradentaries 1 and 3; *Mb.art,* articular pit of meckelian bone; *Prart,* prearticular; *Psym.dp,* parasymphysial dental plate; *de.tusk,* dentary tusk; *f.my,* mylohyoid foramina; *f.r.aut.V, f.r.md.VII* foramina for r.auriculotemporalis V and r.mandibularis of n.facialis; *pr.retr,* processus retroarticularis; *sym.pit,* symphysial pit.

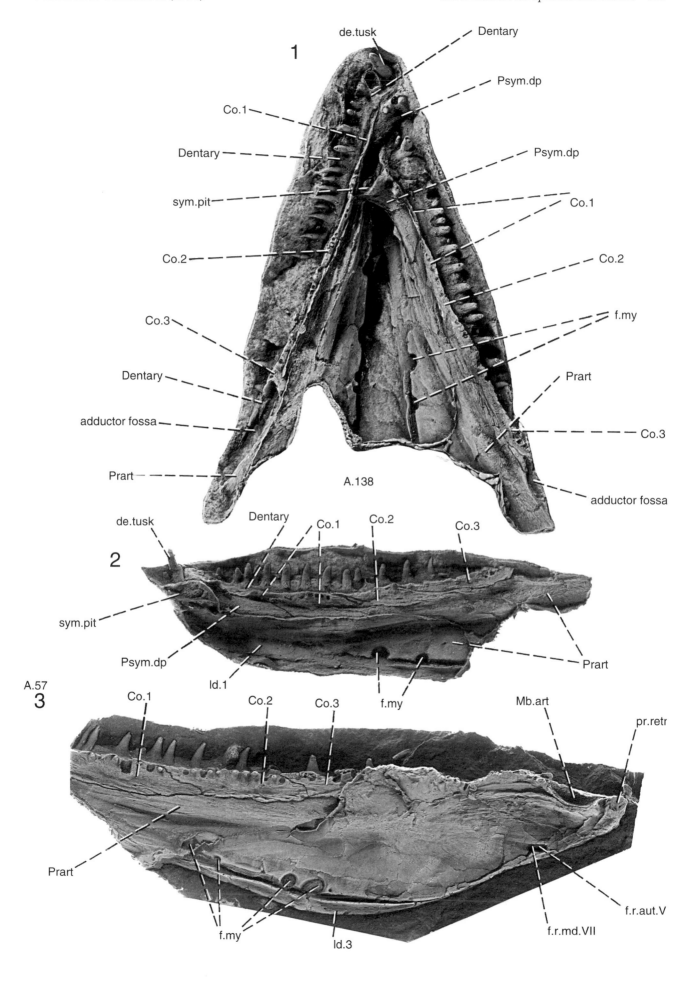

Plate 32

☐1. Lower jaw in lateral view. A. 167, coll. 1932, Stensiö Bjerg. Natural size.

☐2. Anterior part of lower jaw in medial view. A. 83, coll. 1936, Smith Woodward Bjerg. ×2.

☐3. Anterior parts of lower jaw and palate. A. 180, coll. 1951, Sederholm Bjerg (1174 m). Natural size.

Co.1, coronoid 1; *Id.1*, *3*, *4*, infradentaries 1, 3 and 4; *Prart*, prearticular; *Psym.dp*, parasymphysial dental plate; *c.palt*, canals for twigs of r.palatinus; *de.tusk*, dentary tusk; *gr.p-mc*, groove for preoperculo-mandibular sensory canal; *p-mc*, preopercular-mandibular canal; *s.plv.Id 2*, sulcus for vertical pit-line of infradentary 2; *sym.pit*, symphysial pit.

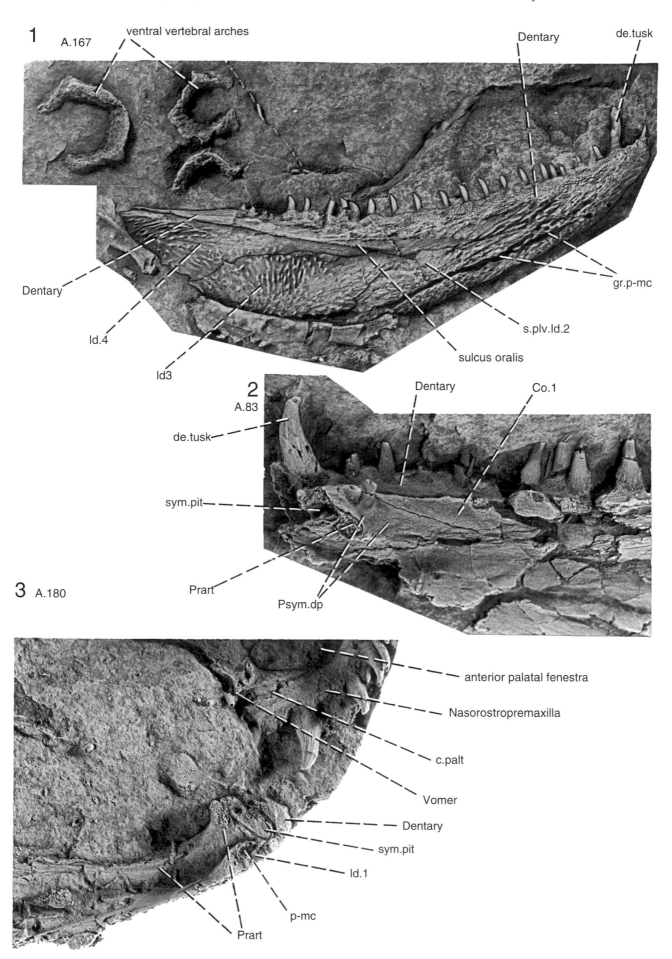

1 A.167 — ventral vertebral arches — Dentary — de.tusk — Dentary — ld.4 — ld3 — sulcus oralis — s.plv.ld.2 — gr.p-mc

2 A.83 — de.tusk — sym.pit — Prart — Psym.dp — Dentary — Co.1

3 A.180 — anterior palatal fenestra — Nasorostropremaxilla — c.palt — Vomer — Dentary — sym.pit — ld.1 — p-mc — Prart

Plate 33

☐1. Lower jaw in lateral view. A. 147, coll. 1949, Smith Woodward Bjerg. Natural size.

☐2. Lower jaw with sensory canal prepared by Säve-Söderbergh. A. 55, same specimen as in Pls. 6:2, 3; 13:1; 26:1. Natural size.

☐3, 4. Anterior part of palate in ventral view and lower jaw in lateral view. A. 163, coll. 1949, Sederholm Bjerg (1174 m). Natural size.

Id.1–Id.4, infradentaries 1–4; *Psym.dp*, parasymphysial dental plate; *de.tusks*, dentary tusks; *gr.p–mc*, groove for preoperculomandibular sensory canal (*p–mc*); *s.plv.Id.2*, sulcus for vertical pit-line of infradentary 2.

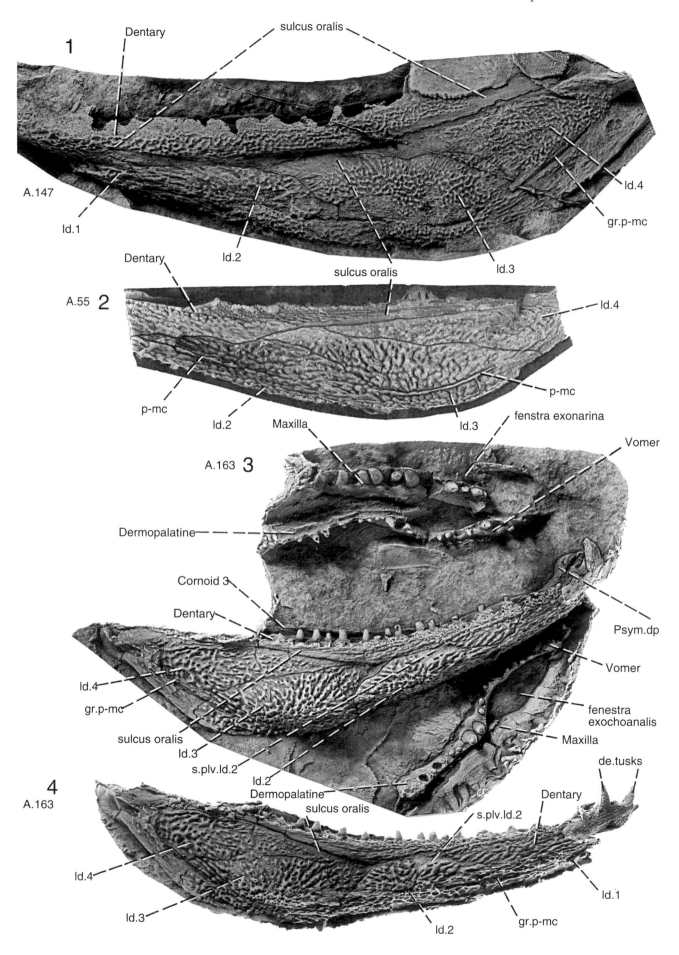

Plate 34

☐1. Cast of postsacral ribs and dorsal part of vertebral column. A. 54, coll. 1934, Sederholm Bjerg (1174 m). $\times \frac{5}{4}$.

☐2. Cast of part of vertebral column of the trunk. A. 140, coll. 1949, Sederholm Bjerg (1174 m). $\times \frac{5}{4}$.

☐3. Small part of A. 140, photographed in alcohol. $\times 3$.

From Jarvik 1952, Fig. 14.

Ido, Interdorsal; *N.a*, neural arch, *R*, postsacral rib; *art.d.R*, area of neural arch articulating with tuberculum of rib; *cap*, capitulum of rib; *f.dr*, *f.vr*, notches for dorsal and ventral roots of spinal nerve; *pozyg*, postzygapophysis; *przyg*, prezygapophysis; *tub*, tuberculum.

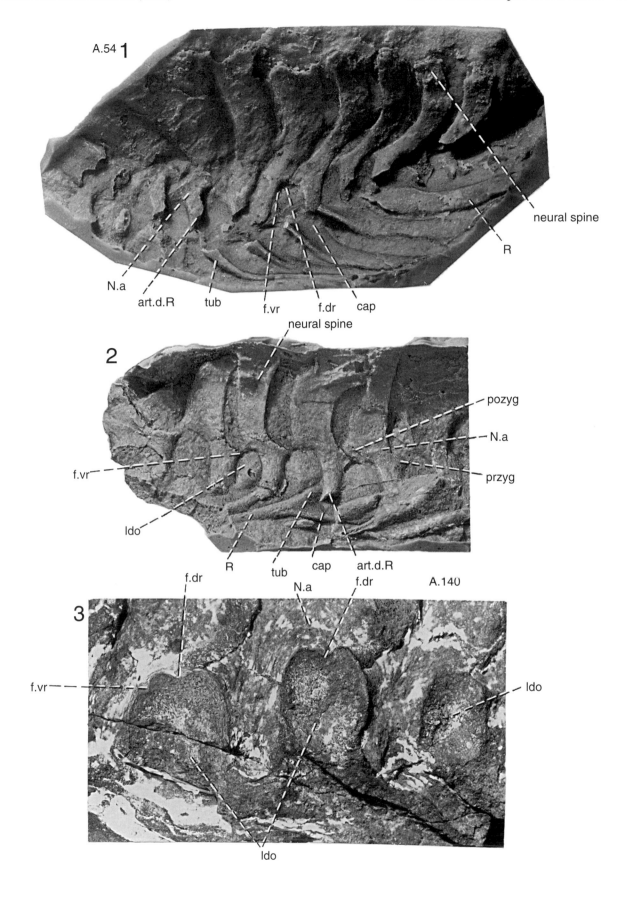

Plate 35

☐1–3. Three ventral vertebral arches in posterior, ventrolateral and ventral views. A. 109, coll. 1948, Celsius Bjerg. Natural size. Same specimen as in Pls. 37, 38, 39:1; 65:1, 2; 66. From Jarvik 1952, Fig. 15.

☐4, 5. Two ventral vertebral arches. A. 92, coll. 1948, Celsius Bjerg, A. 57, coll. 1934, Sederholm Bjerg (1174 m). ×2.

☐6, 7. Latex casts of left ventral vertebral arches in lateral and medial views. Natural size. A. 251, coll. 1955, Celsius Bjerg.

V.a, ventral vertebral arch; *c.n.c*, neural canal; *c.not*, notochordal canal; *gr.a.im*, groove for intermetameric artery; *od.Na*, area overlapped by neural arch; *pap*, parapophysis; *prot*, ventral protuberance.

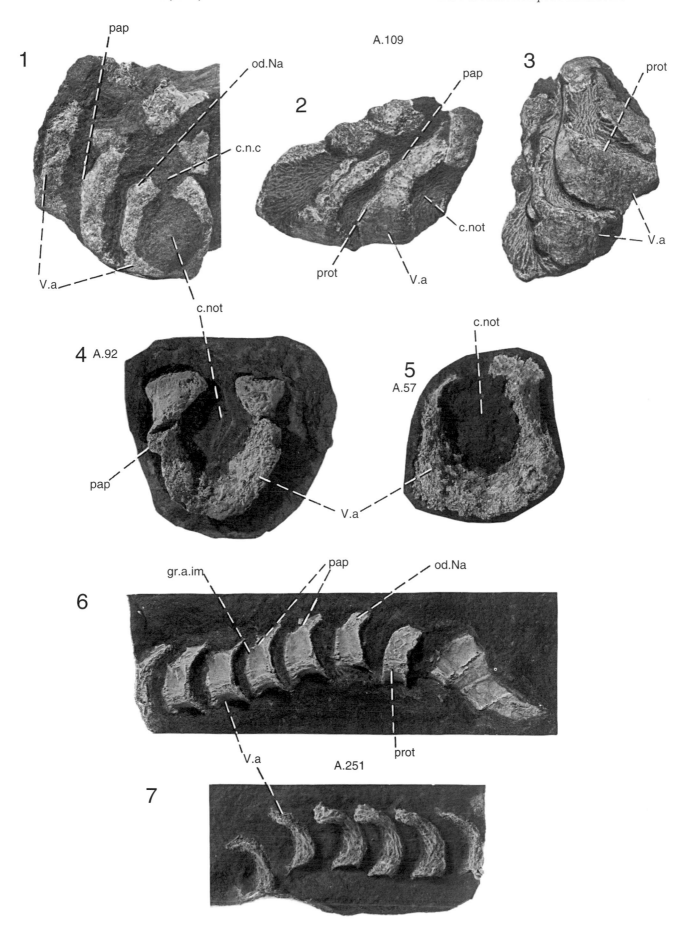

Plate 36

□1, 2. Incomplete tail with counterpart. Natural size. A. 69, coll. 1936, Smith Woodward Bjerg. From Jarvik 1952, Pl. 1.

Dfr, dermal fin rays; *N.a*, neural arch; *Rd, Rv*, dorsal and ventral radials; *V.a*, ventral vertebral arch; *cdf*, dorsal lobe of caudal fin; *c.h*, haemal canal.

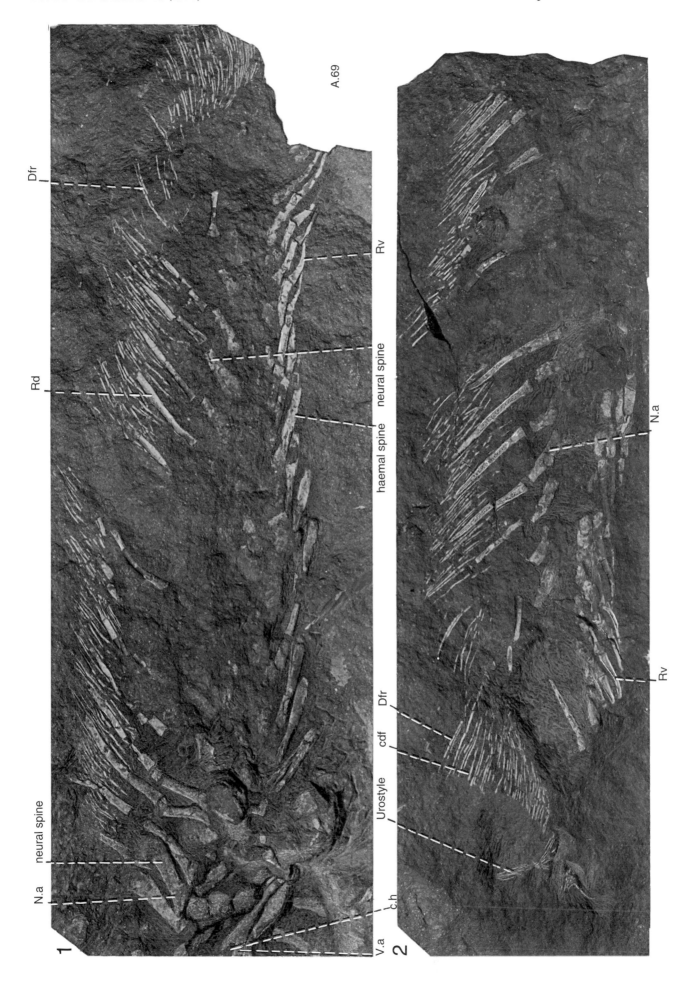

Plate 37

Ventral side of left hindlimb, part of the left pelvic bone and imperfect tail. A. 109, coll. 1948, Celsius Bjerg. Natural size. Same specimen as in Pls. 35:1–3; 38; 39:1; 65:1, 2; 66. From Jarvik 1952, Pl. 2.

Dfr, dermal fin rays; *R.ps*, postsacral rib; *I, V,* digits I and V; *c.sd.li*, canal for supradorsal ligament.

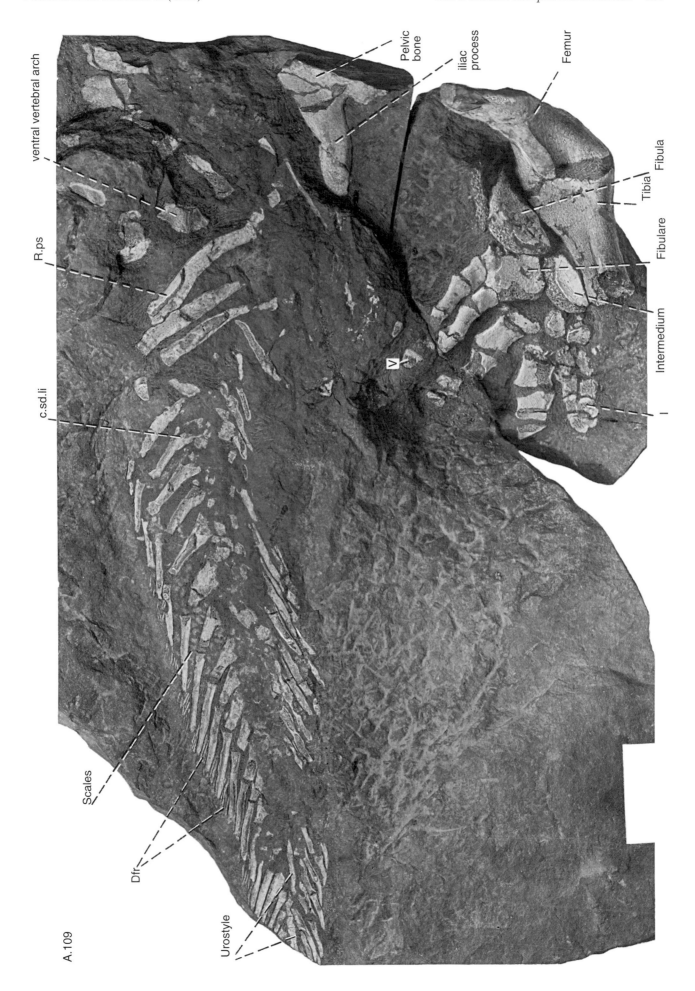

A.109

Plate 38

□1, 2. Imperfect tail with counterpart. $\times^6/_5$. Same specimen as in Pl. 37. From Jarvik 1952, Pl. 3.

Dfr, dermal fin rays; *Rd*, dorsal radial; *R.ps*, postsacral rib; *Rv*, ventral radial; *V.a*, ventral vertebral arch; *c.h*, haemal canal; *c.n.c*, neural canal; *c.sd.li*, canal for supradorsal ligament; *ep.cf*, epichordal lobe of caudal fin.

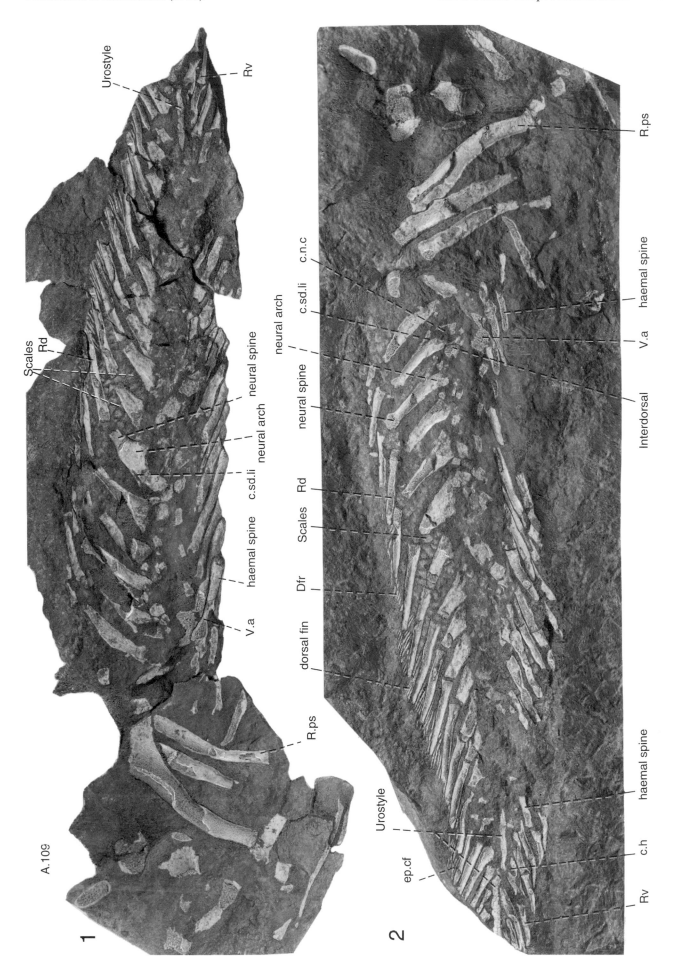

Plate 39

☐1. Posterior part of specimen shown in Pls. 37, 38. ×9/4. From Jarvik 1952, Pl. 4.

☐2. Latex cast of tail. A. 157, coll. 1949, Sederholm Bjerg (1174 m). Natural size. From Jarvik 1959a, Fig. 19 (see also Jarvik 1952, Pl. 6).

Dfr, dermal fin rays; *Rd*, dorsal radial; *Rdi*, distal radial; *Rd+nsp*, dorsal radial fused with neural spine; *Rpx*, proximal radial; *Rv*, ventral radial; *Rv+hsp*, ventral radial fused with haemal spine; *V.a*, ventral vertebral arch; *c.h*, haemal canal; *c.not*, notochordal canal; *ep.cf, hyp.cf*, epi- and hypochordal lobes of caudal fin.

Plate 40

□1, 2. Almost complete tail with counterpart. A. 156, coll. 1949, Sederholm Bjerg (1174 m). Natural size. From Jarvik 1952, Pl. 5.

Dfr, dermal fin rays; *Rd*, dorsal radial; *Rd+nsp*, dorsal radial fused with neural spine; *Rv*, ventral radial; *Rv+hsp*, ventral radial fused with haemal spine; *V.a*, ventral vertebral arch, *c.h*, haemal canal; *ep.cf*, *hyp.cf*, epi- and hypochordal lobes of caudal fin.

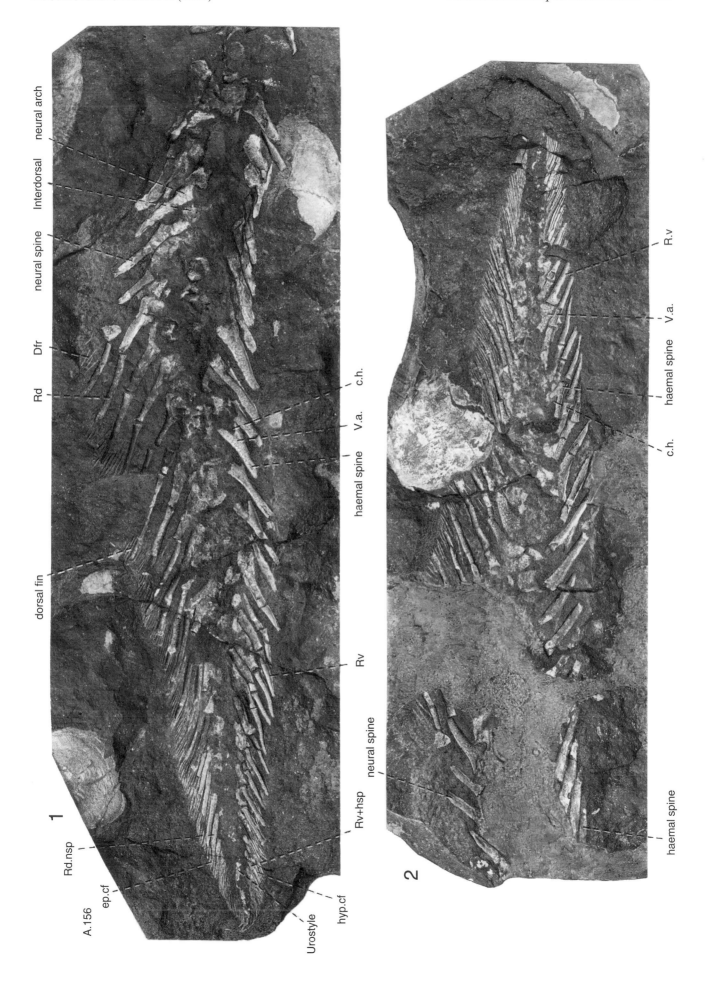

Plate 41

☐1. Thoracal ribs in lateral view.

☐2. The same after removal of the unidentified bone. Specimen No 99, coll. 1929, Celsius Bjerg. $\times ^5/_4$. Same specimen as figured by Stensiö 1931, Pl. 36.

cut.a, cut.v, canals for cutaneous artery and vein.

Plate 42

□1–3. Anterior part of ribcage. (1) left and (2) right sides in lateral view. (3) ventral view. A. 115, coll. 1948, Smith Woodward Bjerg. Natural size. Same specimen as in Pls. 21:5; 45; 46; 51:1; 53; 54; 68:5–7.

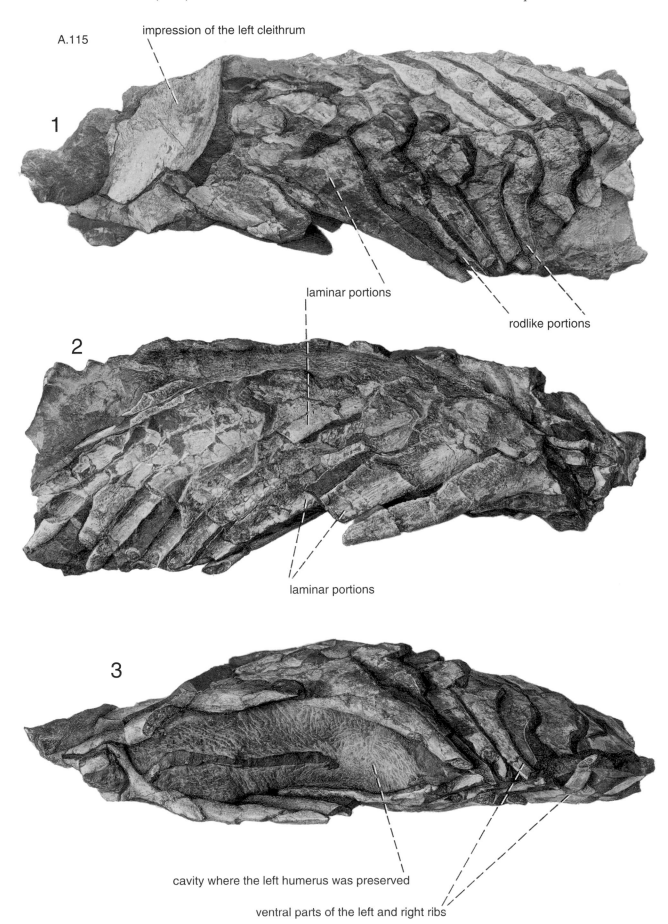

A.115

impression of the left cleithrum

1

laminar portions

rodlike portions

2

laminar portions

3

cavity where the left humerus was preserved

ventral parts of the left and right ribs

Plate 43

Isolated ribs in natural size. 1. A. 152, coll. 1949, Sederholm Bjerg (1174 m). 2. A. 24, coll. 1932, Celsius Bjerg. 3. A. 165, coll. 1949, Sederholm Bjerg (1174 m) 4. A. 25, coll. 1932, Celsius Bjerg. 5. A. 36, coll. 1932, Celsius Bjerg. 6. A. 54, coll. 1954, Sederholm Bjerg (1174 m).

art.cap, articular area of capitulum; *cap*, capitulum of rib; *cut.a, cut.v*, grooves and canals for cutaneous artery and vein; *f.cut*, foramina for cutaneous blood vessels; *og.R.ps*, area overlapping postsacral rib following next behind; *th*, thickening; *tub*, tuberculum.

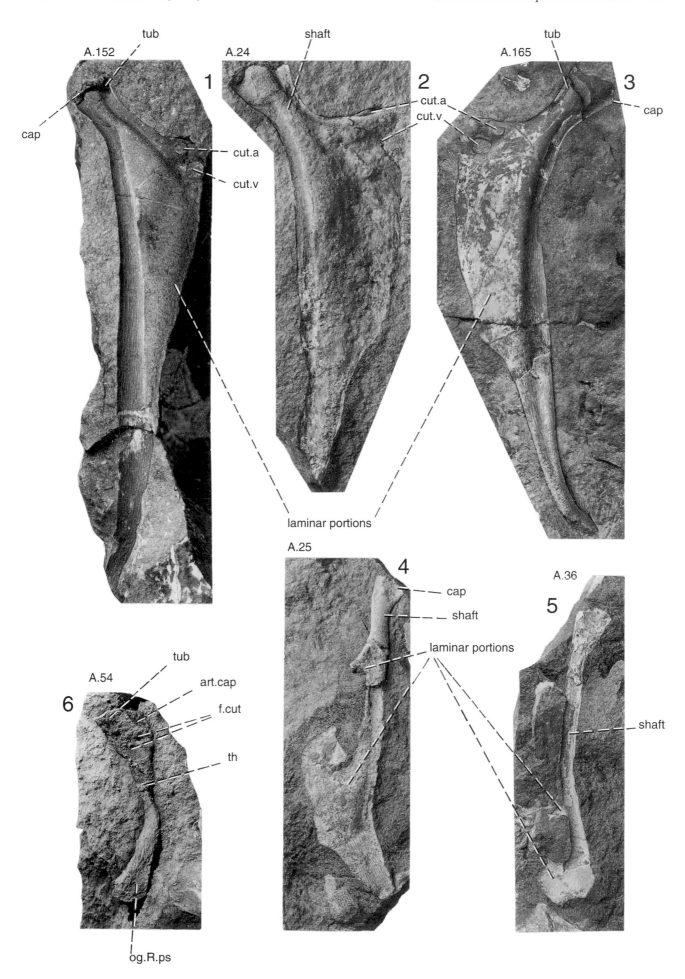

Plate 44

Latex casts of three isolated ribs. 1, 2, A. 173, coll. 1951. Thin cast showing both internal (1) and external (2) sides of rib; 3, A. 235, coll. 1955. External side; 4. A. 172, coll. 1951. External side. All from the horizon at 1174 m, Sederholm Bjerg. Natural size.

cap, capitulum; *cut.a, cut.v*, canals and grooves for cutaneous artery and vein; *f.cut, gr.cut*, foramen and groove for branches of cutaneous blood vessels; *tub*, tuberculum.

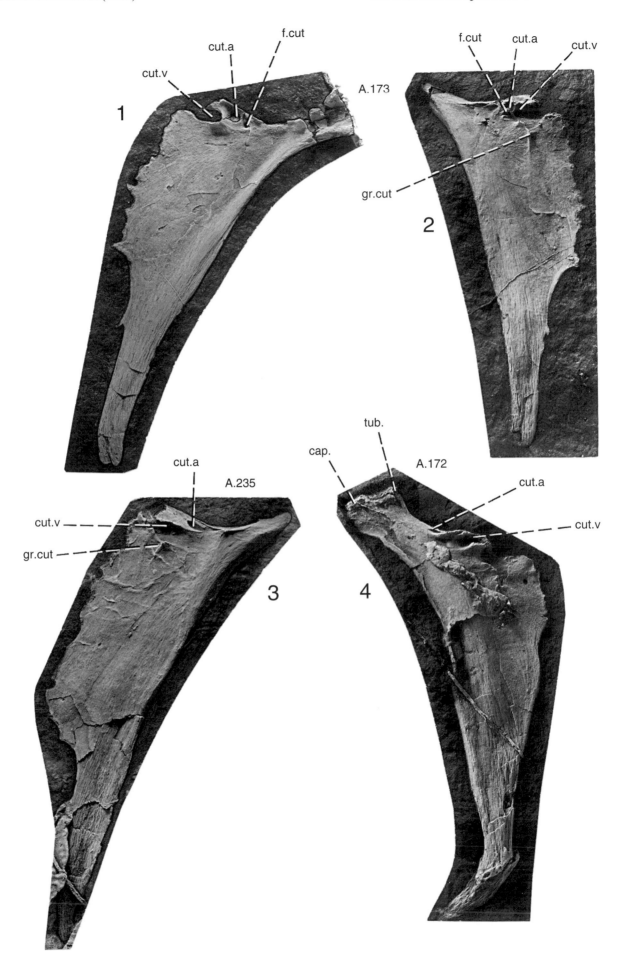

Plate 45

□1–5, Left shoulder girdle in (1) lateral, (2) anteromedial, (3) medial, (4) posterior and (5) anterior views. A. 115, coll. 1948, Smith Woodward Bjerg. Natural size. Same specimen as in Pls. 21:5; 42; 46; 51:1; 53; 54; 68:5–7.

depr, depression on external side of cleithrum with opening of canal; *pr.asc*, ascending process of clavicle.

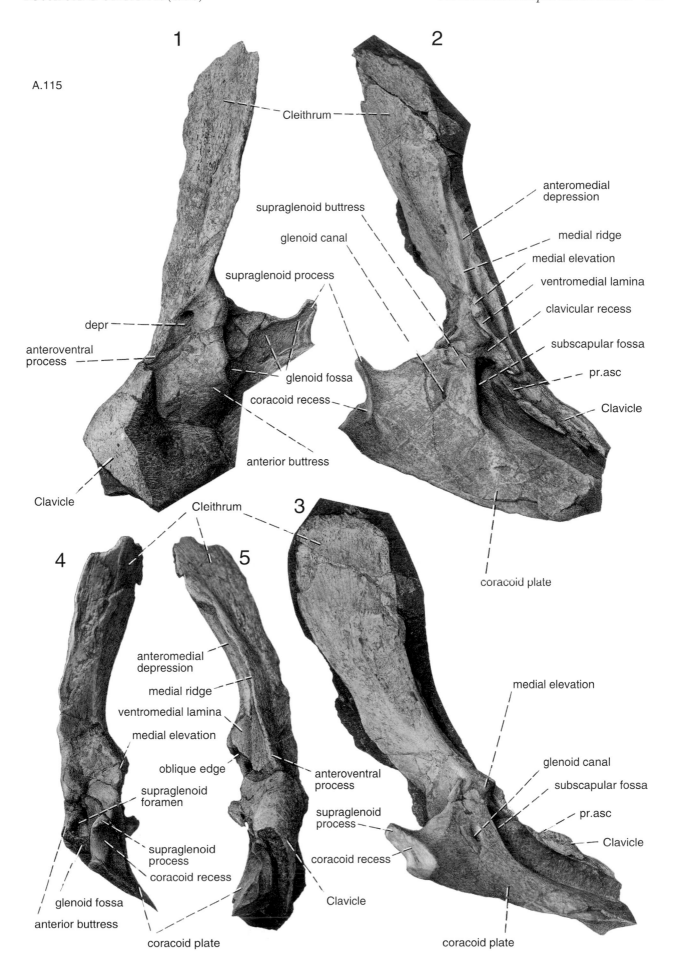

A.115

1

Cleithrum

supraglenoid buttress

glenoid canal

supraglenoid process

depr

anteroventral process

glenoid fossa

coracoid recess

anterior buttress

Clavicle

2

anteromedial depression

medial ridge

medial elevation

ventromedial lamina

clavicular recess

subscapular fossa

pr.asc

Clavicle

coracoid plate

3

Cleithrum

4

Cleithrum

anteromedial depression

medial ridge

ventromedial lamina

medial elevation

oblique edge

supraglenoid foramen

supraglenoid process

coracoid recess

glenoid fossa

anterior buttress

coracoid plate

5

anteroventral process

supraglenoid process

coracoid recess

Clavicle

coracoid plate

medial elevation

glenoid canal

subscapular fossa

pr.asc

Clavicle

coracoid plate

Plate 46

□1–5. A. 115. Right shoulder girdle with humerus and part of skull in various views. Same specimen as in Pls. 21:5; 42; 45; 51:1; 53; 54; 68: 5–7. Natural size.

art.al, art.am, anterolateral and antermedial proximal articular areas of humerus; *art.fo.R*, radial articular fossa; *art.U*, articular area for ulna; *cap*, capitulum of radial fossa; *c.b, c.c, c.d*, canals b, c, and d; *cr.l*, lateral crest; *cr.4–6*, crest between processes 4 and 6; *ent*, entepicondyle; *e.1*, sharp edge; *la.dl, la.dl.ri*, dorsolateral lamina and ridgelike continuation; *la.vl*, ventrolateral lamina; *pr.2, 5, 6*, processes 2, 5, 6 of humerus.

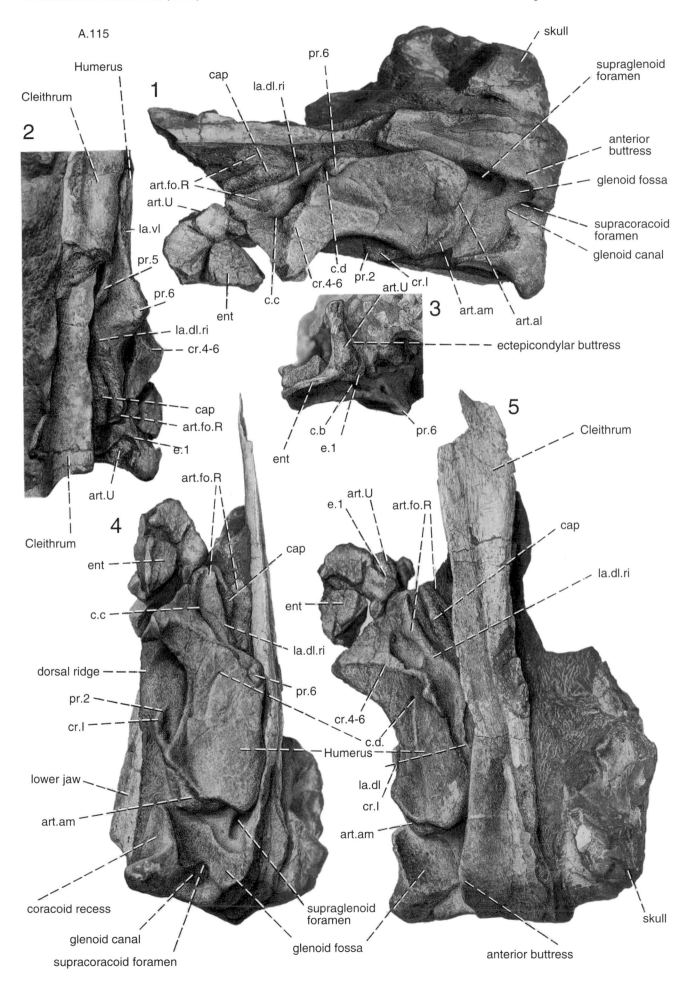

Plate 47

☐1. Ventral part of cleithrum with imperfect endoskeletal shoulder girdle in lateral view. A. 248, coll. 1955, Celsius Bjerg. Natural size.

☐2. Ventral part of the same specimen showing glenoid fossa. ×2.

☐3. Shoulder girdle in medial view. A. 175, coll. 1951, Sederholm Bjerg (1174 m). Natural size.

☐4. Shoulder girdle in lateral view. A. 249, coll. 1955, Celsius Bjerg. Natural size.

depr, depression on external side of cleithrum with opening of a canal.

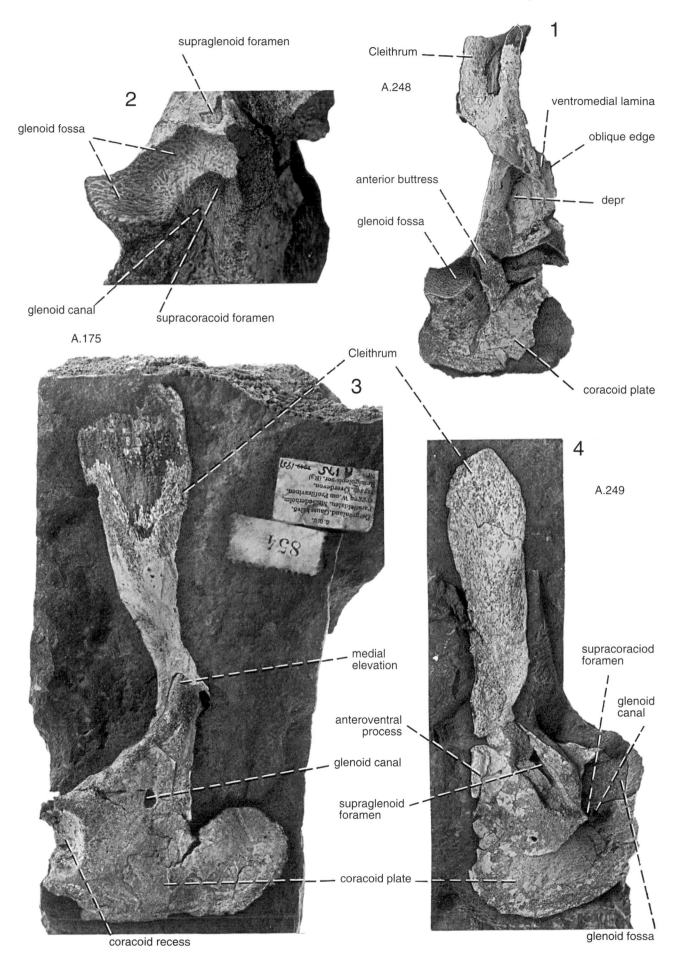

2

supraglenoid foramen

glenoid fossa

glenoid canal

supracoracoid foramen

A.175

1

Cleithrum

A.248

ventromedial lamina

oblique edge

anterior buttress

depr

glenoid fossa

coracoid plate

3

Cleithrum

medial elevation

anteroventral process

glenoid canal

coracoid plate

coracoid recess

4

A.249

supracoraciod foramen

glenoid canal

anteroventral process

supraglenoid foramen

glenoid fossa

Plate 48

□1. Shoulder girdle in lateral view. A. 76, coll. 1936, Remigolepisryg.

□2, 3, ventral part of the same specimen with counterpart. All ×⁵⁄₄.

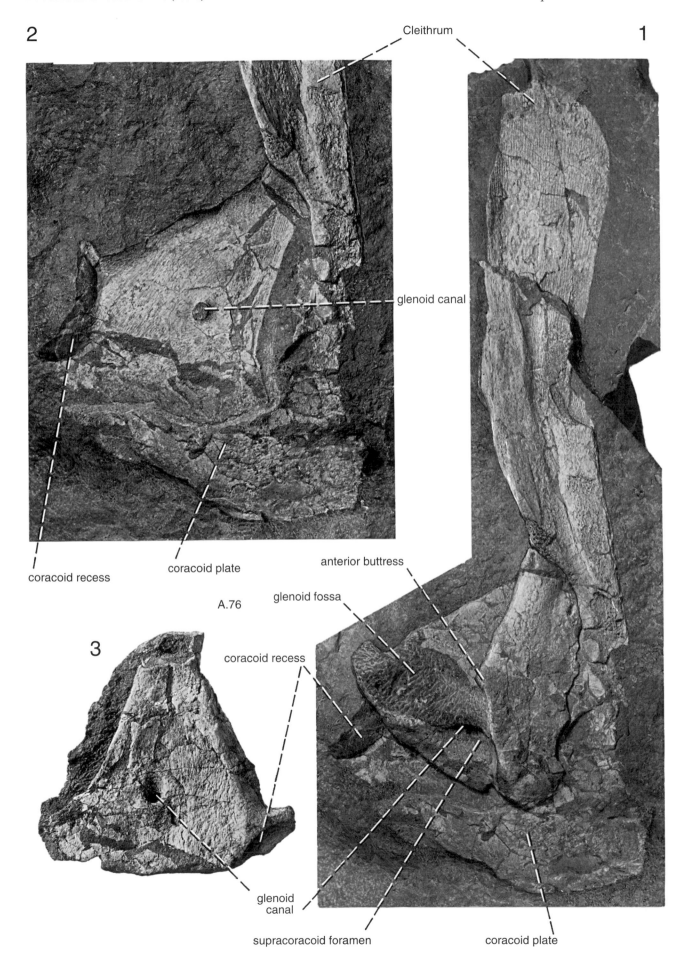

Plate 49

□1–5. Ventral part of cleithrum and parts of the endoskeletal shoulder girdle, showing (1, 3–5) openings of the supracoracoid and supraglenoid canals in the bottom of the subscapular fossa. A. 135, coll. 1949, Sederholm Bjerg (1174 m). All natural size.

depr, depression on external side of cleithrum with opening of a canal.

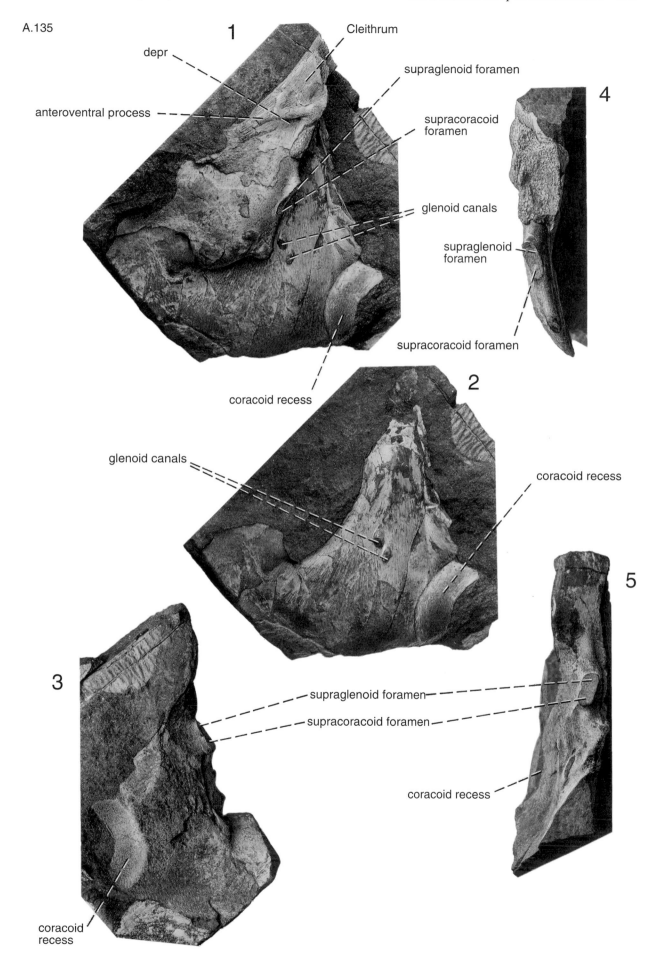

A.135

1

depr

anteroventral process

Cleithrum

supraglenoid foramen

supracoracoid foramen

glenoid canals

coracoid recess

4

supraglenoid foramen

supracoracoid foramen

2

glenoid canals

coracoid recess

5

3

supraglenoid foramen

supracoracoid foramen

coracoid recess

coracoid recess

Plate 50

☐1. Latex cast of isolated rib and parts of shoulder girdle. A. 182, coll. 1951, Sederholm Bjerg (1174 m).

☐2. Isolated clavicle in external view. A. 36, coll. 1932, Stensiö Bjerg.

☐3, 4. Isolated clavicle in ventral and dorsal views. A. 251, coll. 1955, Celsius Bjerg. All natural size.

cut.a, cut.v, canals for cutaneous artery and vein.

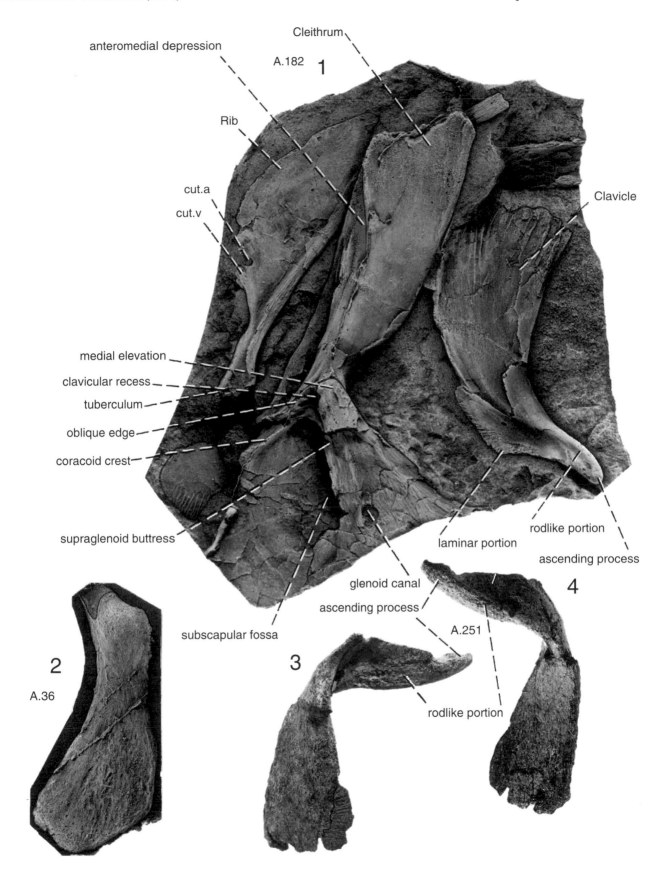

Plate 51

☐1. Latex cast of outside of cleithrum. A. 115. Same specimen as in Pls. 21:5; 42, 45, 46; 53; 54; 68:5–7. $\times^2/_3$.

☐2–5. Isolated clavicle in ventral, dorsal, medial and lateral views. A. 98, coll. 1948, Celsius Bjerg. Natural size.

☐6–7. Isolated clavicle in ventral and lateral views. A. 42, coll. 1934. Sederholm Bjerg. Natural size.

depr, depression on external side of cleithrum with opening of a canal; *orn*, ornamentation.

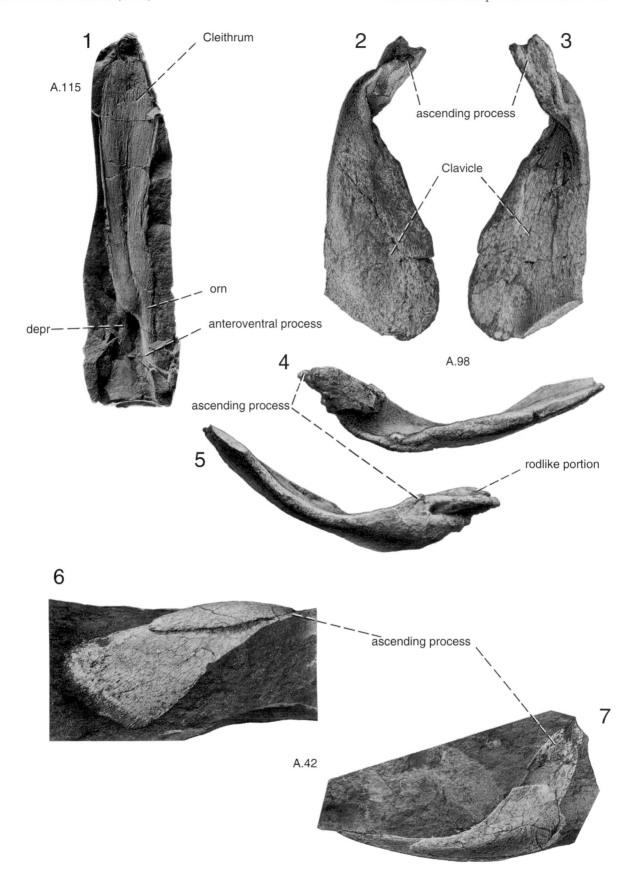

Plate 52

☐1. Interclavicle. Ventral side. A. 39, coll. 1934. Sederholm Bjerg.

☐2. Impression of ventral side of interclavicle A. 146, coll. 1949, Smith Woodward Bjerg.

☐3. Imperfect interclavicle. Dorsal side. A. 79, coll. 1936. Remigolepisryg.

☐4. Imperfect interclavicle. Ventral side. A. 12, coll. 1931, Celsius Bjerg.

All natural size.

od.Clav, area overlapped by clavicle.

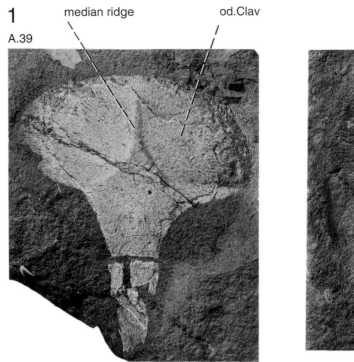

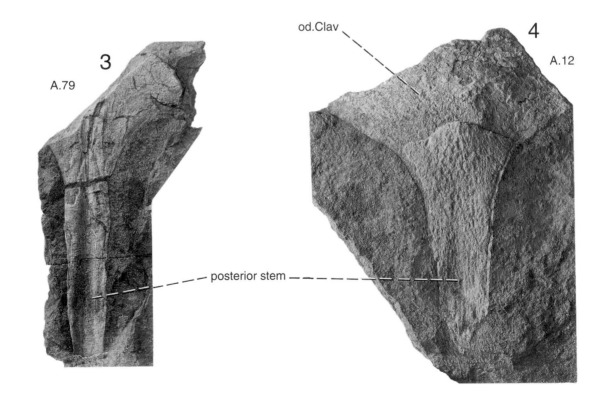

Plate 53

□1–5. Left humerus with radius in (1) lateral, (2) ventral, (3) posterior, (4) anterior and (5) medial views. A. 115. Same specimen as in Fig. 44; Pls. 21:5; 42, 45, 46, 54, 68:5–7. ×$\frac{3}{2}$.

art.al, art.m, anterolateral and anteromedial proximal articular areas; *art.U*, articular area for ulna; *c.b, c.c*, canals b and c; *cr.l*, lateral crest; *cr.1–5, cr.4–6*, crests between processes 1 and 5 and between 4 and 6; *e.l, e.2*, sharp edges; *la.dl, la.vl*, dorsolateral and ventrolateral laminae; *perf*, perforation; *pl.groove*, posterolateral groove of radius; *pr.1, 3–7*, processes 1, 3–7.

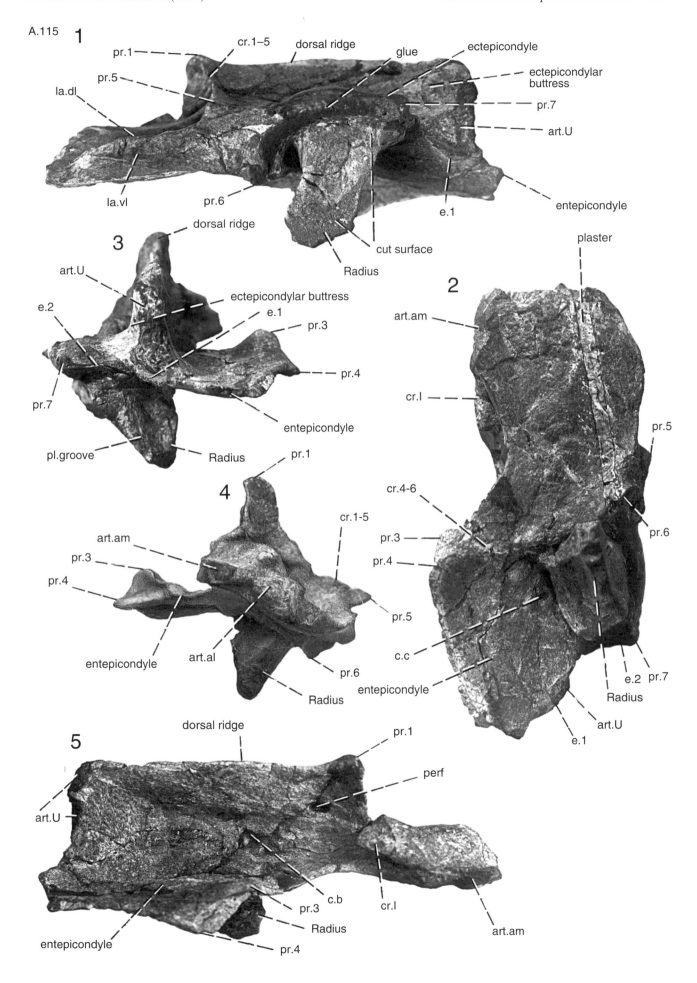

Plate 54

☐1–7. Right radius in (1) medial, (2) lateral, (3) posterolateral views, (4, 5) proximal and distal ends; (6) anteromedial and (7) posteromedial views.

☐8–12. Right ulna. (8) proximal end; (9) medial (10) lateral; (11) anterior and (12) posterior views. All A.115. Same specimen as in Pls. 21:5; 42; 45; 46; 53; 68:5–7. ×$\frac{3}{2}$.

al, am, anterolateral and anteromedial surfaces; *pj.al, pj.am*, anterolateral and anteromedial projections; *pl.groove*, posterolateral groove; *pm*, posteromedial surface; *ri.a, ri.m*, anterior and median ridges; *tri*, triangular area.

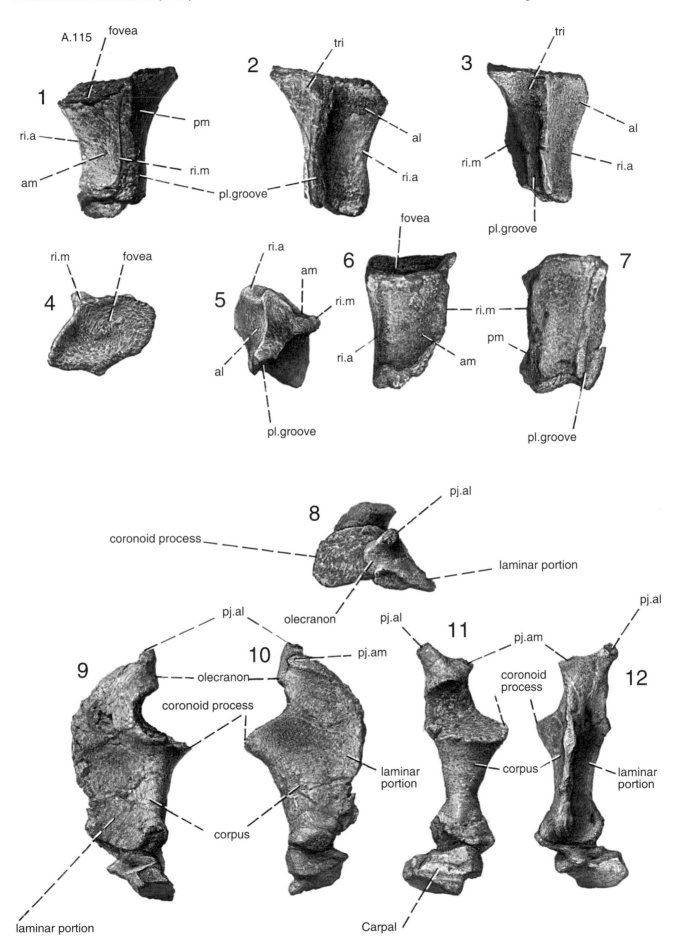

Plate 55

☐1, 2. Almost complete pelvic bone (main part of iliac process missing) in lateral and medial views. A. 93 coll. 1948, Stensiö Bjerg. ×³⁄₂.

pr.a, pr.p, anterior and posterior processes; *to.a*, anterior toungelike projection.

A.93

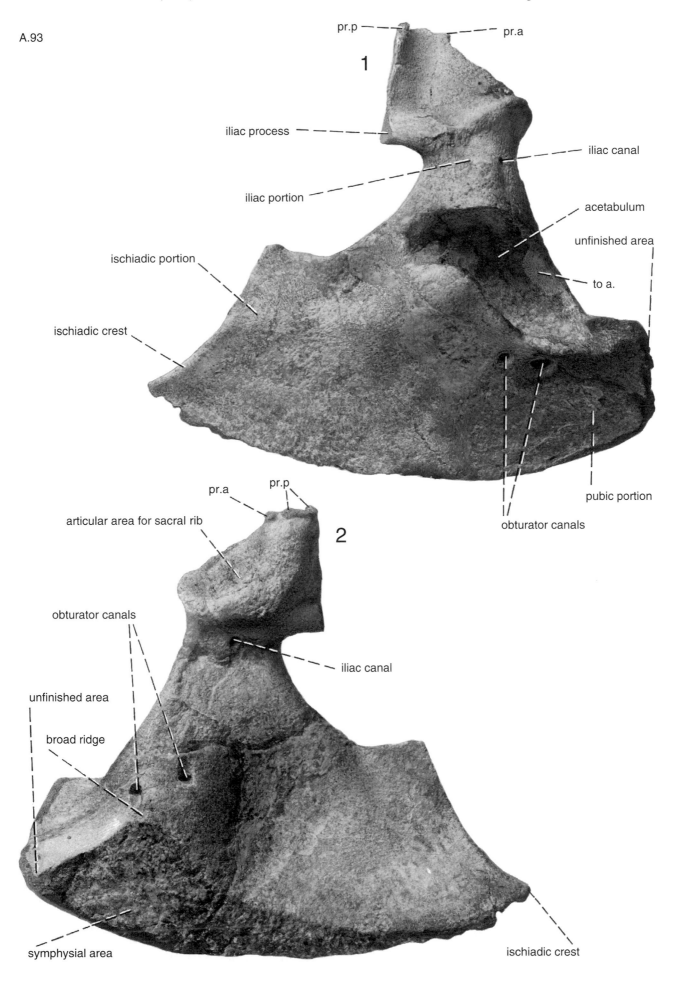

Plate 56

□1, 2. Pelvic bone in lateral and medial views. A. 250, coll. 1955, Celsius Bjerg. $\times^5/_4$.

pr.a, pr.p, anterior and posterior dorsal processes; *to.a, to.p*, anterior and posterior toungelike projections.

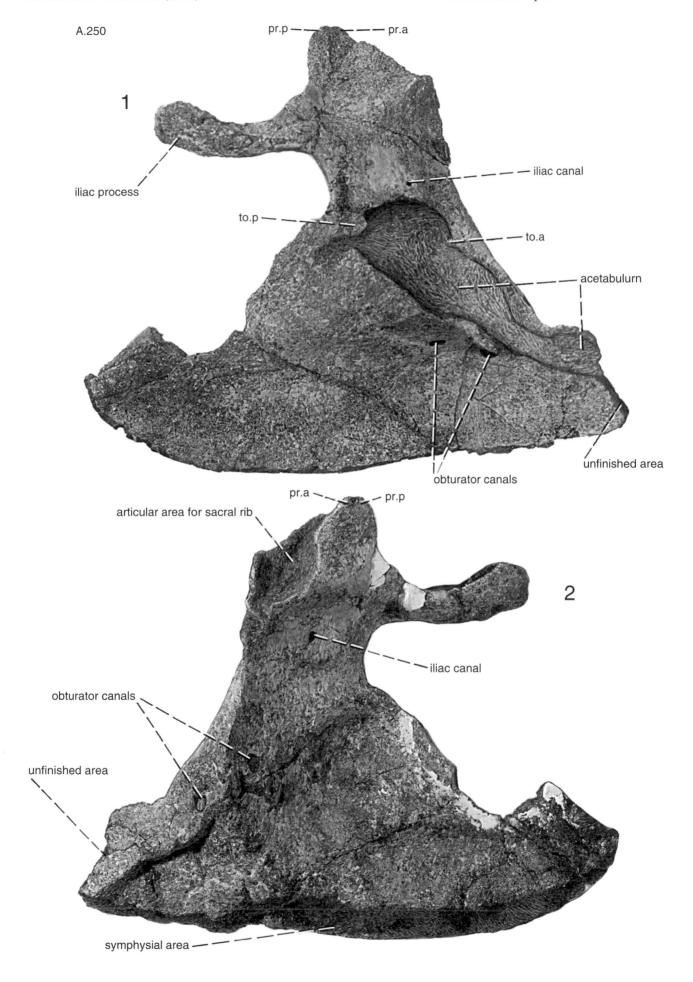

A.250

1

iliac process

pr.p — — pr.a

iliac canal

to.p — — to.a

acetabulurn

obturator canals

unfinished area

articular area for sacral rib

pr.a — — pr.p

2

iliac canal

obturator canals

unfinished area

symphysial area

Plate 57

☐1. Latex cast of pelvic bone. Pubis partly independent. Lateral side. A. 164. Natural size.

☐2. Latex cast of pelvic bone. Lateral side. A. 160. $\times^3/_2$.

Both specimens collected 1949, Sederholm Bjerg (1174 m).

acet.pub, pubic portion of acetabulum; *to.a, to.p*, anterior and posterior tonguelike projections.

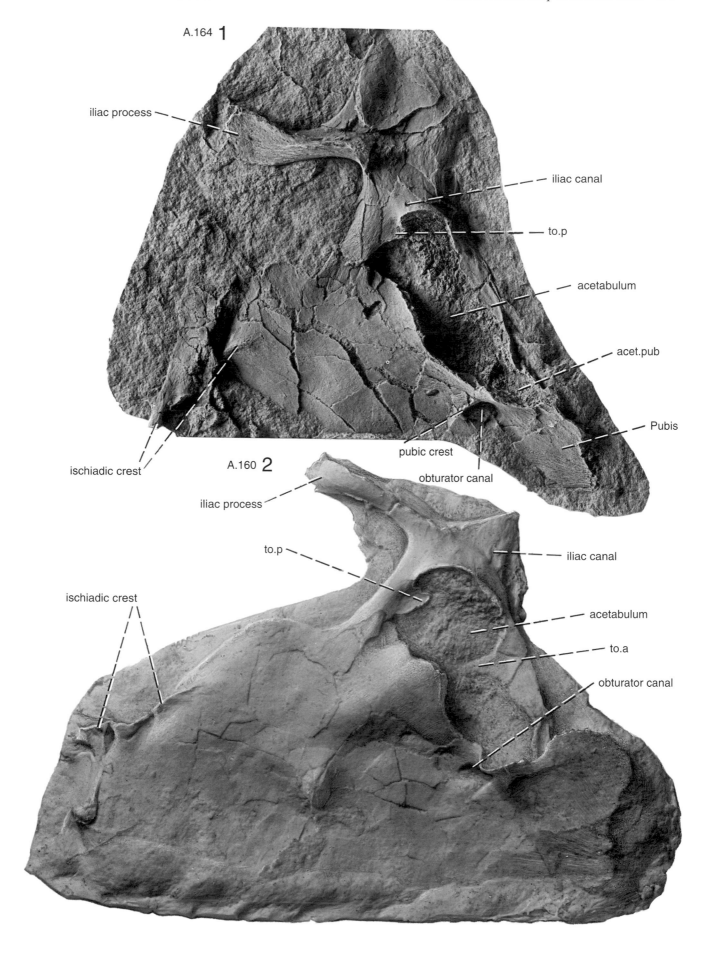

A.164 **1**

iliac process

iliac canal

to.p

acetabulum

acet.pub

Pubis

ischiadic crest

pubic crest

obturator canal

A.160 **2**

iliac process

to.p

iliac canal

ischiadic crest

acetabulum

to.a

obturator canal

Plate 58

☐1. Broken pelvic bone. Medial side. A. 97, coll. 1948, Celsius Bjerg. ×$\frac{5}{4}$.

☐2. Pelvic bone. Medial side. A. 220, coll. 1932, Celsius Bjerg. ×$\frac{3}{2}$.

pr.a, pr.p, anterior and posterior dorsal process.

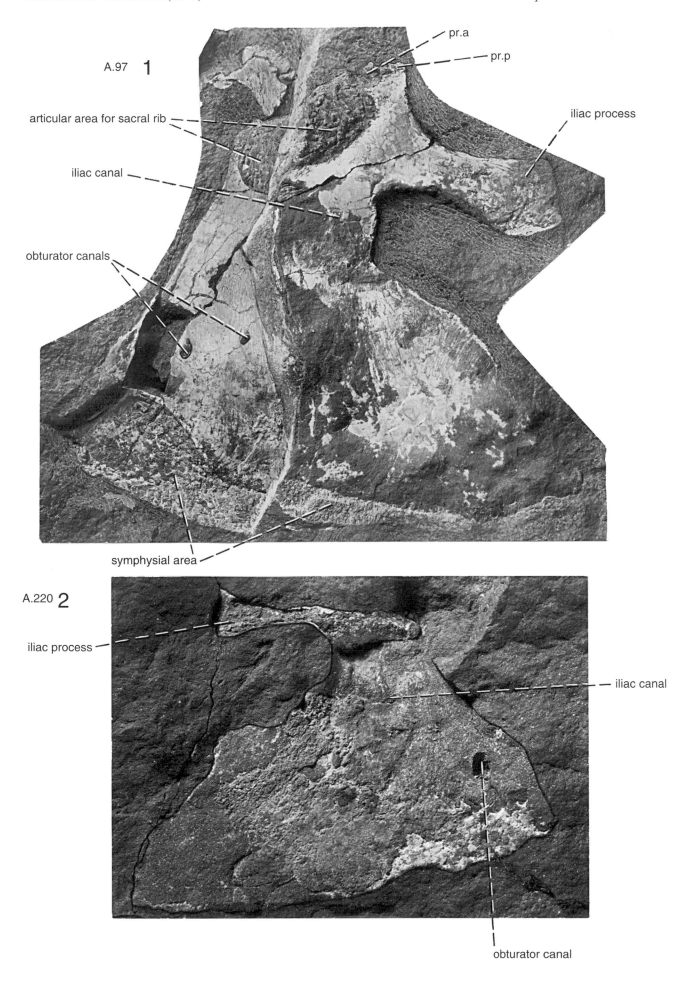

A.97 **1**

pr.a
pr.p
iliac process
articular area for sacral rib
iliac canal
obturator canals
symphysial area

A.220 **2**

iliac process
iliac canal
obturator canal

Plate 59

☐1–4. Pelvic bone shown in Pl. 55 in (1) ventrolateral, (2) dorsal, (3) posterior and (4) anterior views. ×³⁄₂.

pr.a, pr.p, anterior and posterior dorsal processes.

A.93

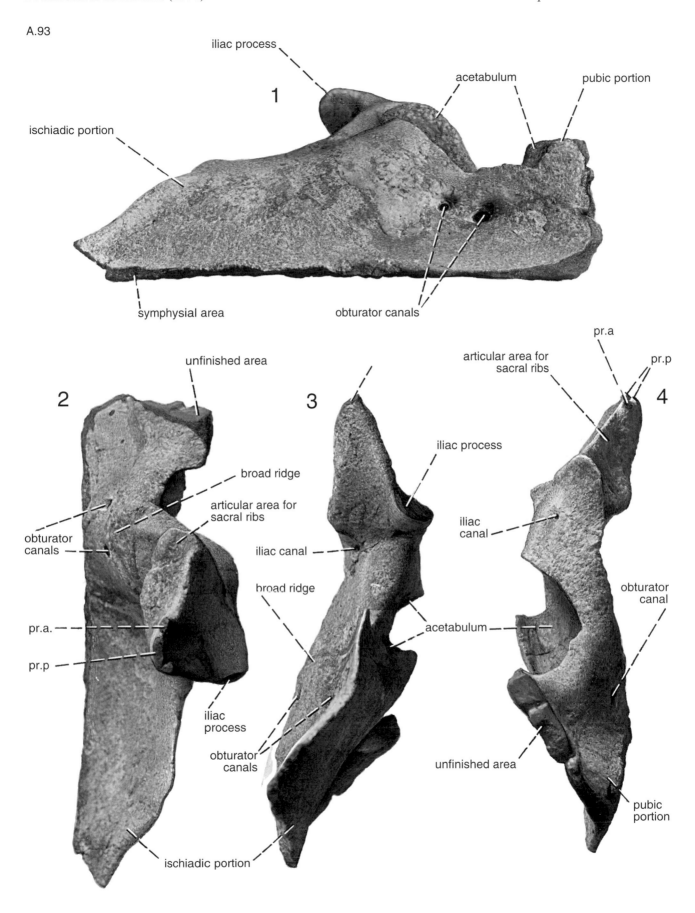

Plate 60

□1. Impression of part of lateral side of pelvic bone. A. 155, coll. 1949, Sederholm Bjerg (1174 m). $\times \frac{3}{2}$.

□2. Loose piece of the same bone showing part of the medial side. $\times 3$.

to.a, *to.p*, anterior and posterior tonguelike projections.

A.155

Plate 61

☐1–4. Both pelvic bones with isolated rib. A. 132, coll. 1949, Sederholm Bjerg (1174 m). Natural size.

☐1. Left pelvic bone in lateral view.

☐2. Part of the same in dorsolateral view.

☐3. Iliac portion of right pelvic bone in lateral view.

☐4. Parts of both pelvic bones in ventral view.

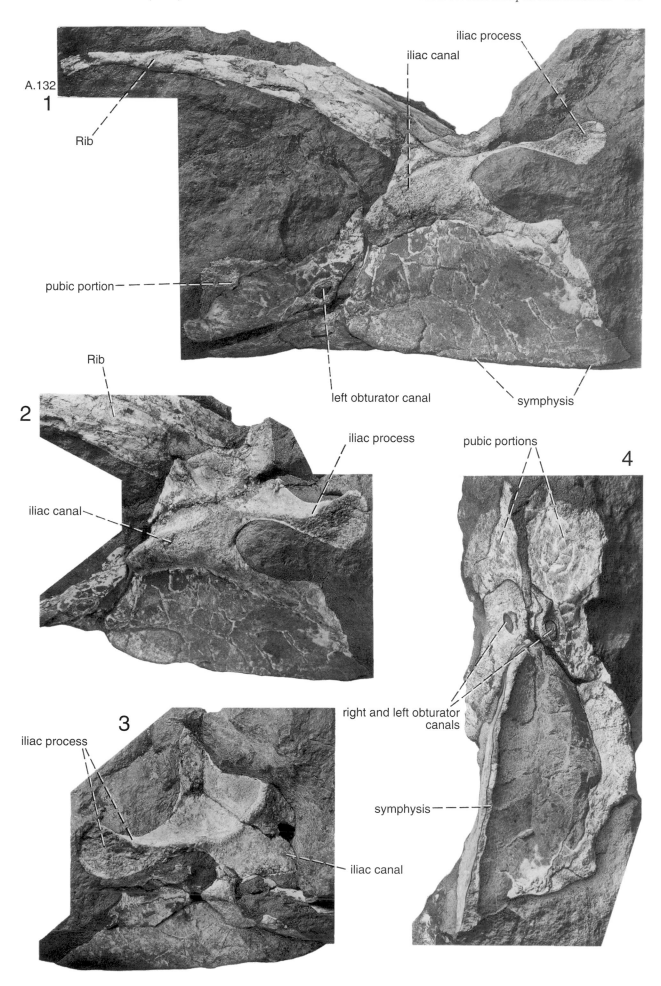

Plate 62

□1–2. Parts of the pelvic bones and left hindlimb. Pubis partly independent. A. 166, coll. 1949, Sederholm Bjerg (1174 m). All natural size. Same specimen as in Fig. 52 and Pls. 63, 64.

□1. Pelvic bones with part of foot in dorsal view.

□2. Right pelvic bone in medial view.

□3. Pelvic girdle with partly independent pubis. A. 163, coll. 1949, Sederholm Bjerg (1174 m). ×⅔.

IV, V, digits IV and V.

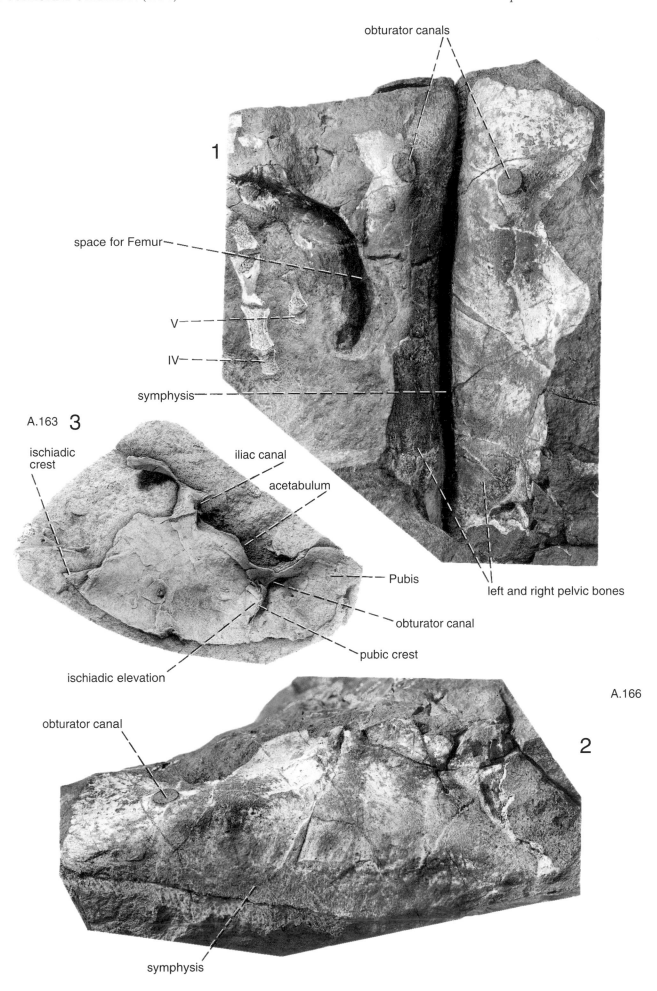

obturator canals

space for Femur

V

IV

symphysis

A.163 **3**

ischiadic
crest

iliac canal

acetabulum

Pubis

obturator canal

pubic crest

ischiadic elevation

left and right pelvic bones

A.166

2

obturator canal

symphysis

Plate 63

□1, 2. Left hindlimb showing mainly dorsal side of foot before and after preparation. A. 166. Latex cast. In Fig. 52.

□3. Latex cast of ventral side of foot. A. 166. Same specimen as in Pls. 62:1, 2; 64. Natural size.

Idigit, interdigital structure: I–V, digits I–V.

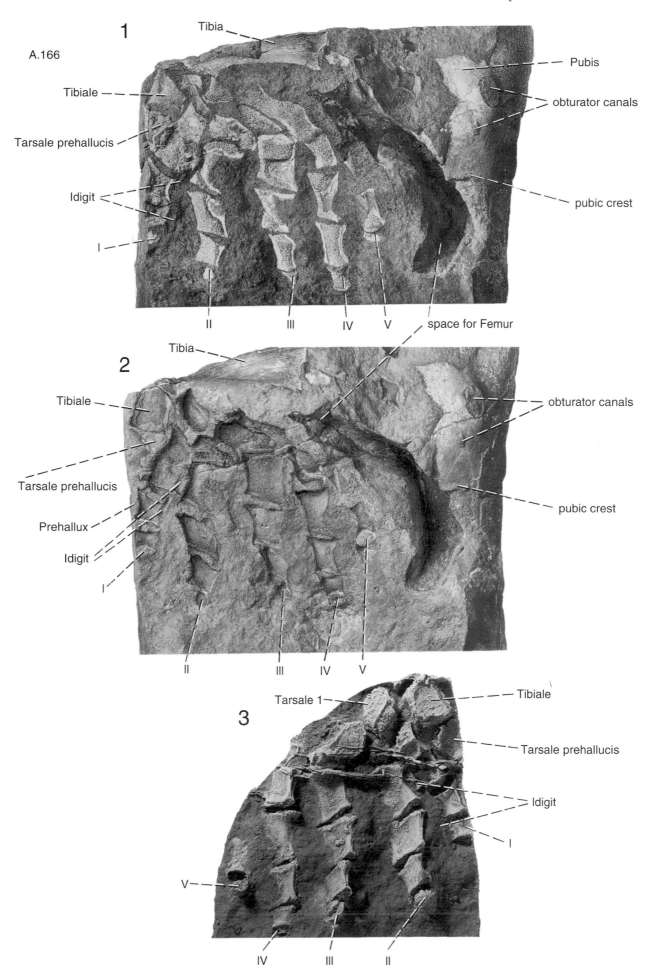

Plate 64

☐1, 2. Counterpart of dorsal side of foot shown in Pl. 63:1, before and after preparation.

☐3. Latex cast of the foot shown in 2.

☐4, 5. Proximal part of the same foot with latex cast. All natural size.

Idigit, interdigital structure; *art.im*, area of tibia articulating with intermedium. *I–V*, digits I–V.

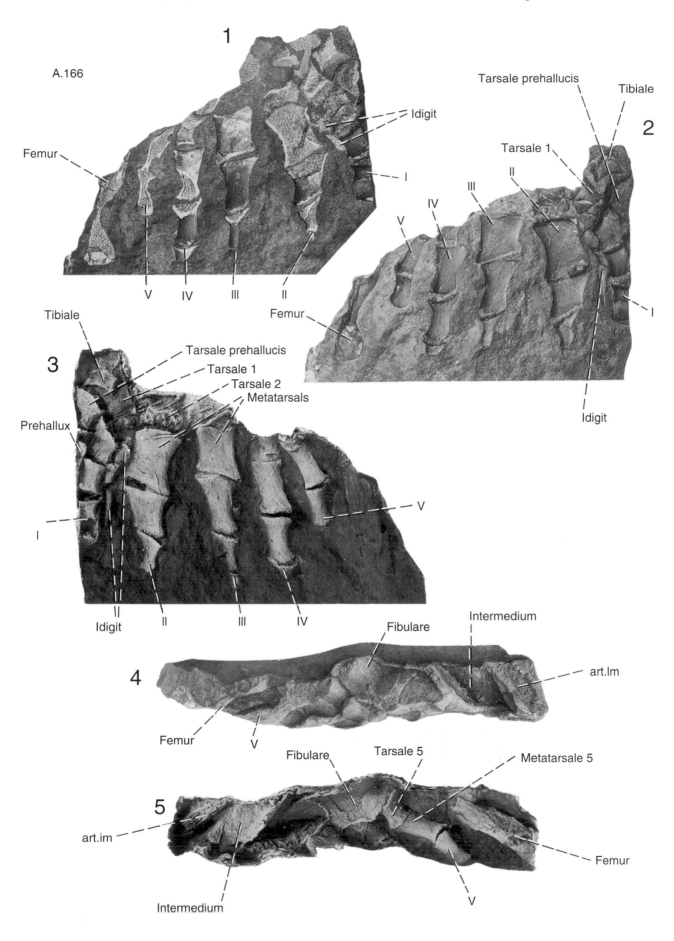

Plate 65

□1, 2. Same hindlimb and fibula as shown in Pl. 37. $\times^3/_2$.

□3. Isolated tibia. A. 112, coll. 1948, Smith Woodward Bjerg. Natural size.

art.Fe, art.Im, art.Tib, areas of tibia articulating with femur, intermedium and tibiale; *art.Pmin*, area of fibula articulating with postminimus; *la.Ti, th.Ti*, laminar and thickened portions of tibia. *I–V*, digits I–V.

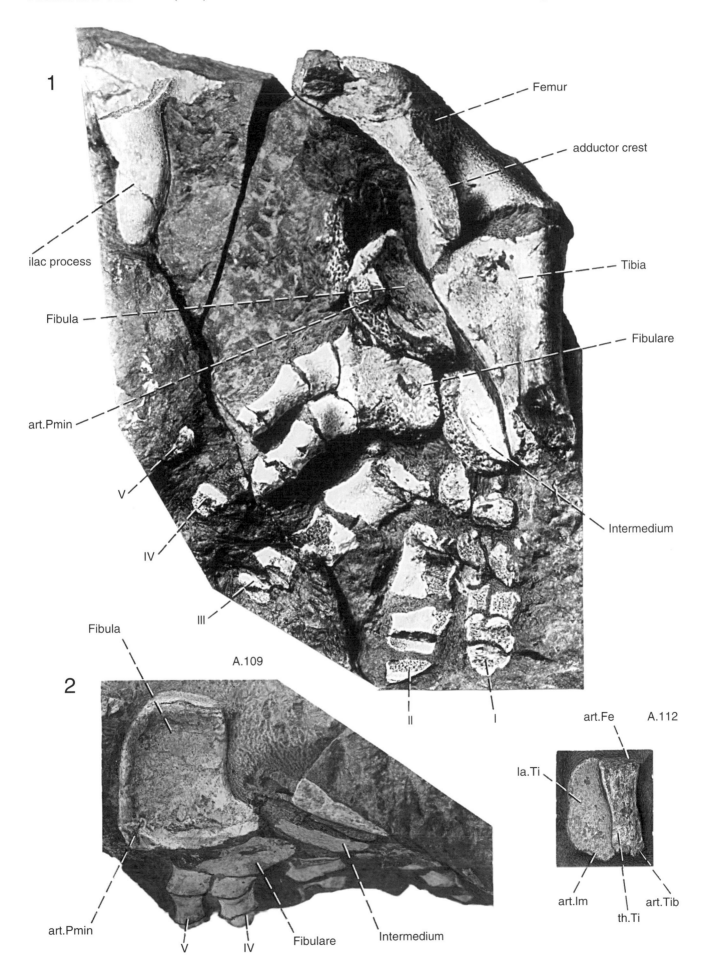

1

Femur

adductor crest

ilac process

Tibia

Fibula

Fibulare

art.Pmin

V

Intermedium

IV

III

A.109

II

I

2

Fibula

art.Pmin

V   IV   Fibulare   Intermedium

art.Fe   A.112

la.Ti

art.Im   art.Tib

th.Ti

Plate 66

☐1–4. Femur and tibia of specimen shown in Pl. 65 in (1) lateral, (2) medial (3) posterior and (4) anterior views. ×$\frac{3}{2}$.

art.Im, art.Fi, art.Tib, areas articulating with intermedium, fibula and tibiale; *c.Fe.1, c.Fe.2,* femoral canals 1 and 2; *depr.Ti,* depression of tibia; *la.Ti,* laminar portion of tibia; *ri.Fi, ri.Ti,* fibular and tibial ridges; *th.Ti,* thickened portion of tibia; *tr.4,* fourth trochanter.

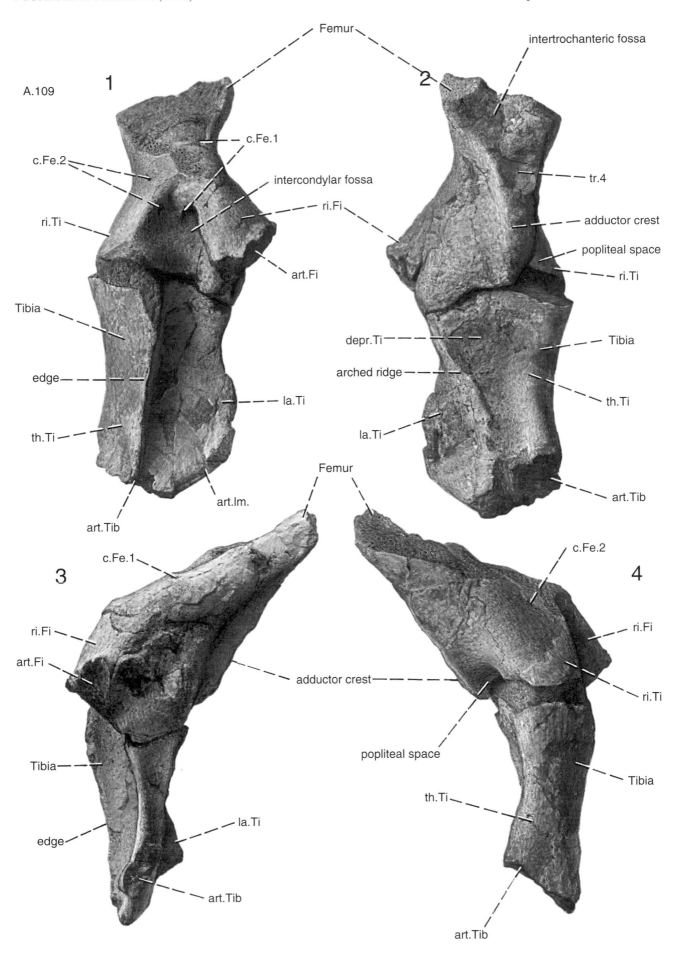

Plate 67

□1–6. Isolated femur. A. 77, coll. 1936, Sederholm Bjerg in (1) lateral, (2) medial, (3) posterior, (4) anterior, (5) proximal and (6) distal views. ×$\frac{3}{2}$.

art.Fi, art.Ti, articular areas for fibula and tibia; *c.Fe.1*, femoral canal 1; *prox.art*, proximal articular area; *ri.Fi, ri.Ti*, fibular and tibial ridges or condyles.

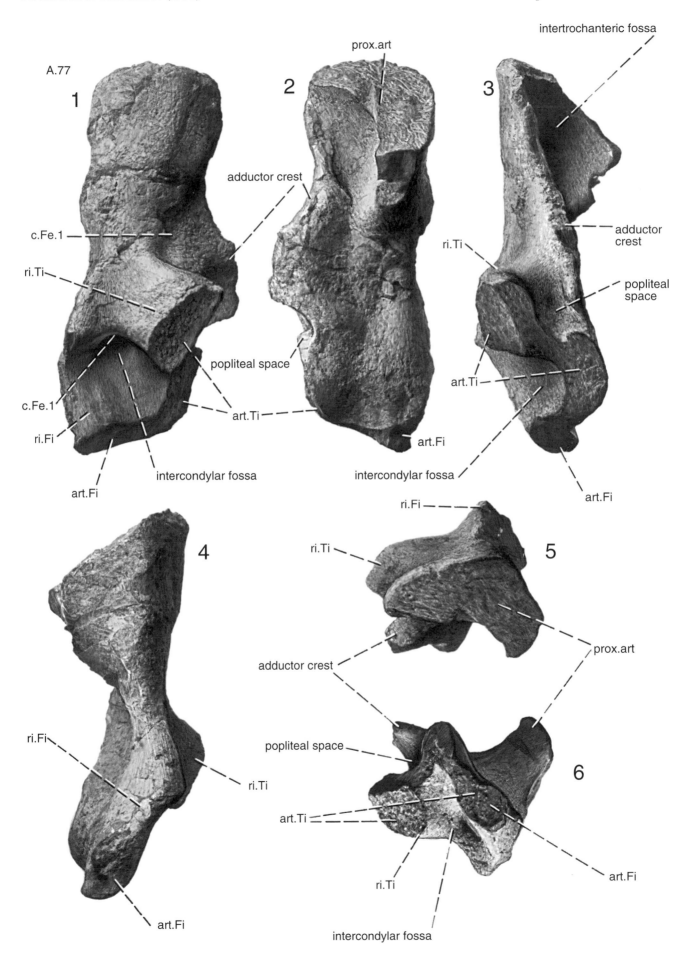

Plate 68

☐1–4. Compressed isolated right femur. A. 65, coll. 1934, Sederholm Bjerg in (1) lateral, (2) medial, (3) ventral and (4) dorsal views. $\times^{3}/_{2}$.

☐5–7. Imperfect left femur and tibia in (5) internal, (6) external and (7) ventral views. A. 115. Same specimen as in Pls. 21:5; 42, 45, 46, 53. $\times^{3}/_{2}$.

art.Fi, art.Ti, articular areas for fibula and tibia; *c.Fe.1, c.Fe.2*, femoral canals 1 and 2; *la.Ti*, laminar portion of tibia; *ri.Fi, ri.Ti*, fibular and tibial ridges or condyles.

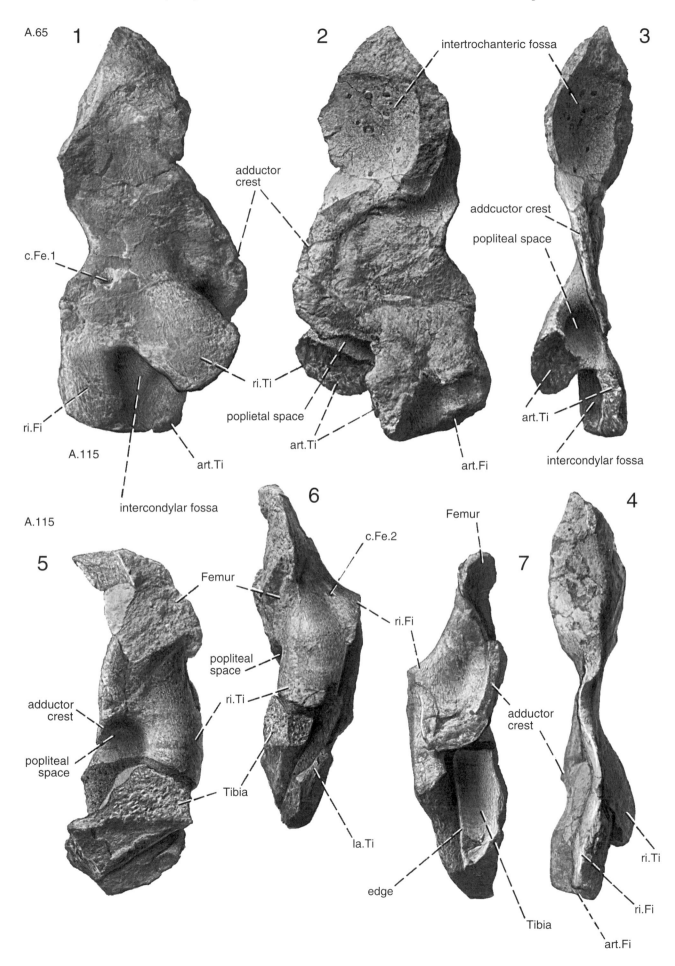